RAND NATIONAL DEFENSE RESEARCH INSTITUTE

Enhancing U.S. Coast Guard Metrics

Scott Savitz, Henry H. Willis, Aaron C. Davenport, Martina Melliand,
William Sasser, Elizabeth Tencza, Dulani Woods

Prepared for the U.S. Coast Guard Atlantic Area

For more information on this publication, visit www.rand.org/t/rr1173

Library of Congress Control Number: 2015955493
ISBN: 978-0-8330-9156-7

Published by the RAND Corporation, Santa Monica, Calif.

Preface

This report is part of an effort to identify ways that the U.S. Coast Guard could improve how it measures its operational performance. Selecting the right operational-level metrics and using them effectively can help the Coast Guard with resource allocation, risk management, and external communications. The authors present an analysis of the elements of Coast Guard missions and the Coast Guard's existing metrics and propose potential metrics and a framework for using them. These findings are currently being considered by the Coast Guard and have not been adopted as Coast Guard doctrine.

This research was sponsored by the U.S. Coast Guard Atlantic Area. It was conducted within the International Security and Defense Policy Center and the RAND Homeland Security and Defense Center of the RAND National Defense Research Institute, a federally funded research and development center sponsored by the Office of the Secretary of Defense, the Joint Staff, the Unified Combatant Commands, the Navy, the Marine Corps, the defense agencies, and the defense Intelligence Community.

The RAND Homeland Security and Defense Center (HSDC) conducts analysis to prepare and protect communities and critical infrastructure from natural disasters and terrorism. Center projects examine a wide range of risk-management problems, including coastal and border security, emergency preparedness and response, defense support to civil authorities, transportation security, domestic intelligence, and technology acquisition. Center clients include the U.S. Department of Homeland Security, the U.S. Department of Defense, the U.S. Department of Justice, and other organizations charged with security and disaster preparedness, response, and recovery.

HSDC is a joint center of two research divisions: RAND Justice, Infrastructure, and Environment and the RAND National Security Research Division. RAND Justice, Infrastructure, and Environment is dedicated to improving policy and decisionmaking in a wide range of policy domains, including civil and criminal justice, infrastructure protection and homeland security, transportation and energy policy, and environmental and natural resource policy. The RAND National Security Research Division conducts research and analysis for all national security sponsors other than the U.S. Air Force and the Army. The division includes the National Defense Research Institute, a federally funded research and development center whose sponsors include the Office of the Secretary of Defense, the Joint Staff, the Unified Combatant Commands, the defense agencies, and the U.S. Department of the Navy. The National Security Research Division also conducts research for the U.S. intelligence community and the ministries of defense of U.S. allies and partners.

For more information on the RAND International Security and Defense Policy Center, see http://www.rand.org/nsrd/ndri/centers/isdp.html or contact the Center director (contact information provided on the web page).

Questions or comments about this report should be sent to the project leader, Scott Savitz (ssavitz@rand.org). For more information about the Homeland Security and Defense Center, see http://www.rand.org/hsdc or contact the director at hsdc@rand.org.

Contents

Figures

Tables

Summary

To effectively measure its operational performance, the U.S. Coast Guard needs to have both the right metrics for this purpose and a framework for understanding them. The Coast Guard's Atlantic Area asked the RAND Corporation to conduct the research documented here to help improve on existing metrics and develop such a framework.

The purpose of this report is to help the Coast Guard improve its ability to measure its performance. The report is structured around three key questions:

- What aspects of operational performance should the Coast Guard aim to measure?
- How well does the Coast Guard measure operational performance?
- How could the Coast Guard measure operational performance better?

We answered the first question by developing logic models that described the Coast Guard's missions, drawing on Coast Guard documents and experts. To answer the second question, we analyzed the Coast Guard's existing metrics. We addressed the third question by developing potential metrics, evaluating them, and outlining a framework for reviewing the values of both existing and potential metrics. We discuss our methodology and findings in greater detail below.

Developing Logic Models

As indicated above, we ascertained what aspects of operational performance the Coast Guard should aim to measure by developing logic models to characterize the Coast Guard's missions, based on Coast Guard sources. Based on Coast Guard Publication 3-0,[1] we structured our analysis around the 11 statutory missions identified in the Homeland Security Act of 2002, under the three broad headings of safety, security, and stewardship.

We developed a logic model structure, as shown in Figure S.1. *Strategic outcomes* relate to the three categories of missions mentioned above: safety, security, and stewardship. These may be closely aligned with the Coast Guard's overarching priorities, such as those enumer-

[1] Coast Guard Publication 3-0, *Operations*, Washington, D.C.: U.S. Coast Guard, February 2012.

Figure S.1
Structure of the Logic Models

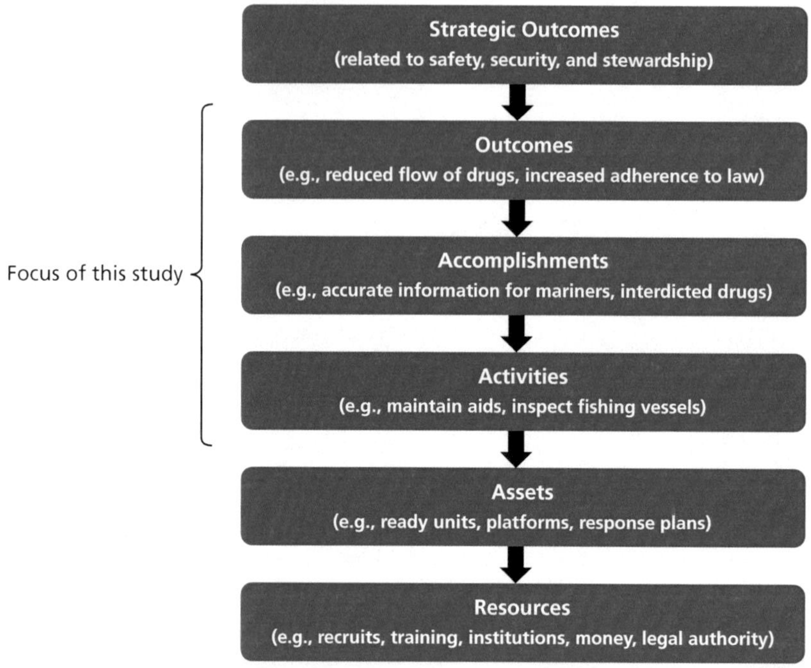

RAND *RR1173-S.1*

ated in the *Coast Guard Western Hemisphere Strategy*,[2] *Arctic Strategy*,[3] and others.[4] Meeting strategic outcomes requires the fulfillment of mission-specific outcomes. *Outcomes* represent the culmination of a mission, such as lives saved or reduced risk from terrorism. Outcomes encompass more-specific aspects of a mission, termed *accomplishments*. Fulfilling accomplishments requires *activities*, such as patrols or inspections. These are conducted with *assets*, such as helicopters; small boats; command, control, communications, computers, intelligence, surveillance, and reconnaissance networks; units of personnel; and documentation. Those assets ultimately draw on *resources*, the Coast Guard's raw materials, such as recruits, funding, and legal authority.

Given that this research was intended to analyze operational-level metrics, we developed the three levels of the logic models that are relevant to operations: activities, accomplishments, and outcomes. For brevity, we collectively refer to these as the *elements* of the logic models.

We developed the 11 logic models (one for each statutory mission) by researching numerous Coast Guard documents and interviewing dozens of subject-matter experts at the headquarters, area, and district levels of the Coast Guard, then garnering their feedback at multiple stages of the analysis.

---

[2] U.S. Coast Guard (USCG), *Western Hemisphere Strategy*, Washington, D.C., September 2014d.

[3] USCG, *Arctic Strategy*, Washington, D.C., May 2013c.

[4] For example, strategic outcomes could be linked with the *Western Hemisphere Strategy*'s three priorities: combating networks, securing borders, and safeguarding commerce. They could also be linked with the goals of USCG, *Command, Control, Communication, Computers, and Information Technology (C4&IT) Strategic Plan FY13–17*, November 2012b. On a geographically specific basis, strategic outcomes could be linked to *Arctic Strategy*'s strategic objectives, namely improving awareness, modernizing governance, and broadening partnerships (USCG, 2013c).

Analyzing Existing Metrics

To assess how well the Coast Guard was measuring operational performance, we analyzed existing metrics, which we gathered from documents and experts. Ultimately, we found 157 metrics, spanning all 11 missions, which were being disseminated among multiple Coast Guard commands. Next, we mapped each of the existing metrics to the elements of the logic models. This enabled us to determine which elements were currently being measured and which were not.[5] Overall, the number of existing metrics (157) was three-quarters as large as the number of elements of the 11 logic models (213). However, these metrics primarily measured less than one-third of the elements of the logic models; some elements of the logic models were measured by multiple metrics, and others by none. This is expected, since the metrics long predate the logic models. Not surprisingly, existing metrics tended to measure the most visible, tangible elements of missions, excluding the others. For example, elements dealing with partnerships, intelligence, investigations, and disseminating information tended not to be measured by existing metrics.

In addition, we evaluated how well the elements were being measured. Based on prior RAND work,[6] we applied three criteria to each existing metric:

- **validity**—the extent to which the metric captures the outcome, accomplishment, or activity being assessed
- **reliability**—how consistently measurements can be made
- **feasibility**—how easily the measurement can be made.

Most of the existing metrics have high reliability and feasibility, reflecting the fact that they are relatively objective measures that can be collected without imposing substantial burdens on operators. However, about one-half of the metrics have either medium or low validity with respect to the elements of the logic models. This is not a critique of these metrics' creators—they created these metrics before RAND developed the logic models in late 2014—but an indication that there is sometimes limited alignment between existing metrics and the elements of the logic models that were identified from Coast Guard documents and interviews with Coast Guard experts.

Identifying Potential Metrics

The third and final question guiding this study was how the Coast Guard can better measure its operational performance. To address this, we identified a set of potential metrics, as well as a framework that would structure the use of both existing and potential metrics. We discuss the potential metrics immediately below, followed by the framework.

The potential metrics we identified served two main purposes. Most of them are intended to measure elements of the logic models that are not currently being measured; we identified at

---

[5] Some of these metrics may also be used for purposes other than measuring operational performance. We did not evaluate their utility in other contexts.

[6] See Stephanie Young, Henry H. Willis, Melinda Moore, and Jeffrey Engstrom, *Measuring Cooperative Biological Engagement Program (CBEP) Performance: Capacities, Capabilities, and Sustainability Enablers for Biorisk Management and Biosurveillance*, Santa Monica, Calif.: RAND Corporation, RR-660-OSD, 2014.

least one potential metric for every element without existing metrics. Many of these elements of the logic models center on the less-visible parts of missions—for example, those relating to interagency or international partnerships.[7] The potential metrics that measure these can complement existing metrics to provide a more comprehensive characterization of operational performance.

The remaining potential metrics are intended to improve on existing metrics in some way. For example, some leverage existing metrics that measured absolute numbers of items but use an appropriate denominator to generate a percentage or a rate. Others include qualifying phrases that adjust the scope of a metric, while still others take a single metric and divide it into multiple closely related metrics.

We recommend implementing only a fraction of the potential metrics, either to complement or replace existing metrics. The set of potential metrics should be viewed as a menu from which the Coast Guard can select, based on a combination of internal priorities and our evaluation of the metrics. To facilitate the Coast Guard's selections, we have categorized the potential metrics based on how they rated with respect to our three criteria: validity, reliability, and feasibility. The 107 metrics that rated as high in all three categories, and the 51 that had high validity with at least medium reliability and feasibility, are the most likely candidates for Coast Guard implementation; however, we have included others for potential consideration.

Composing a Framework

Finally, we constructed a framework for reviewing the values associated with metrics in a structured fashion. Key challenges for multiple levels of Coast Guard leadership are to be able to readily visualize the implications of large numbers of metrics and to be able to track down the sources for any deficiencies in the values of metrics. To that end, we have developed a framework that, given additional supporting analysis, could be used to review multiple levels of metrics in the context of decisionmaking.

Concluding Remarks

As noted at the outset, this study analyzed three main questions:

- *What aspects of operational performance should the Coast Guard aim to measure?* To answer this, we characterized the Coast Guard's missions using logic models that were based on Coast Guard documents and expert insights.
- *How well does the Coast Guard measure operational performance?* We addressed this by mapping existing metrics to the logic models and evaluating them according to three criteria: validity (the extent to which the metric measures what it is intended to measure), reliability (the consistency with which measurements can be made), and feasibility (how easily measurements can be made). We found that 48 percent of existing metrics have

[7] A number of high-profile Coast Guard documents highlight the criticality of partnerships. For example, USCG, 2014d, is replete with discussion of partnerships and capacity-building. USCG, 2012b, cites improvement of information sharing with partners as its first goal. Likewise, USCG, 2013c, lists broadening partnerships as one of its three strategic objectives.

high validity, while 74 percent have high reliability, and 82 percent have high feasibility; only 34 percent rate highly on all three criteria. Moreover, only 29 percent of the elements of the logic models were captured by existing metrics.

- *How could the Coast Guard measure operational performance better?* We answered this question by identifying sets of potential metrics for Coast Guard consideration. We identified potential metrics that measured the 71 percent of the elements of the logic models that were not covered by existing metrics. In addition, we identified potential metrics that had higher levels of validity, reliability, and/or feasibility than existing metrics. We grouped them based on those evaluations for Coast Guard consideration. Finally, we developed a framework that enables Coast Guard leadership to review the values of select metrics associated with multiple levels of the logic models, assuming that additional analysis takes place to enable clear modeling of relationships among the values of individual metrics.

These findings provide options for the Coast Guard to consider as part of its overall efforts to improve operational planning and evaluation. The Coast Guard can select additional metrics to complement or supersede existing metrics, while also enabling senior leadership to readily review the values of metrics by structuring them in the proposed framework.

The Coast Guard can coordinate with other components of the Department of Homeland Security, as well as with the Department of Defense, National Oceanic and Atmospheric Administration, Environmental Protection Agency, and other agencies, to align operational metrics in a way that facilitates communication and decisionmaking.

In addition, the Coast Guard can begin to examine metrics relating to the asset and strategic outcome levels of the logic model, aligning the metrics with the ones discussed in this report. Metrics used to measure mission-support efforts can also be evaluated and aligned. The Coast Guard can also delve more deeply into the proposed framework, evaluating the appropriate values to associate with particular metrics and modeling the relationships among the values of particular metrics.

Acknowledgments

We would like to thank the many people throughout the U.S. Coast Guard who were instrumental in helping us to do this research. These included numerous subject-matter experts and stakeholders at Coast Guard Headquarters, Atlantic Area, Pacific Area, all nine districts, Force Readiness Command, and the Intelligence Coordination Center.

We would also like to thank our immediate sponsors at Atlantic Area—Jacqueline Earle, T. J. Williams, LCDR Blair Sweigart, CAPT William Csisar III, and Joseph DiRenzo—for taking the time to share their thoughts, to help us reach out to other experts, and to review our work as it progressed.

We would also like to explicitly thank our reviewers, VADM (ret.) Robert Parker and Paul DeLuca, for providing direct feedback regarding the manuscript. Seth Jones, Michael McNerney, and Christopher Chivvis supervised this research, helping to shape and improve our work. Several of our colleagues provided useful critiques and comments as the research progressed, including RADM (ret.) Michael McMahon and Paul Davis.

Our sincere apologies if we have missed anyone. Naturally, any errors in the report are the responsibility of the authors.

CHAPTER ONE

Introduction

The U.S. Coast Guard (USCG) has an unusually diverse mission space for a relatively small service.[1] It is responsible for performing 11 statutory missions, such as icebreaking, drug interdiction, and fisheries enforcement, over domestic areas ranging from Maine to Guam; it also contributes to numerous overseas operations. Given these responsibilities, the Coast Guard needs to continually find ways to make the most effective use of its very limited resources.

In light of this, it is critical for the Coast Guard to be able to measure its operational performance with respect to particular missions. This helps the Coast Guard recognize where it could potentially reallocate resources to alleviate shortfalls or how it can manage risk when it reallocates resources to accommodate contingencies. Moreover, by measuring its performance well, the Coast Guard can more effectively communicate with external stakeholders—including Congress, the White House, and the American public—about how well it is doing in particular areas and about the logic behind its choices for allocation and investment.

To effectively measure its operational performance, the Coast Guard needs to have both the right metrics for this purpose and a framework for understanding them. The Coast Guard's Atlantic Area (LANTAREA) asked the RAND Corporation to conduct a study to help improve on existing metrics and develop such a framework. Please note that the Coast Guard is currently considering our findings and has not adopted them as Coast Guard doctrine.

Purpose and Structure of This Report

The purpose of this report is to help the Coast Guard improve its ability to measure its performance. The document is structured around three key questions:

- What aspects of operational performance should the Coast Guard aim to measure?
- How well does the Coast Guard measure operational performance?
- How could the Coast Guard measure operational performance better?

[1] The Coast Guard has only about one-eighth as many active-duty personnel as the U.S. Navy (USN) or Air Force, and about one-fifteenth of the budget of either. The entire Coast Guard budget for 2013—$10.5 billion—is less than the up-front cost of the USS *Gerald R. Ford* aircraft carrier. U.S. Air Force, "United States Air Force Fiscal Year 2015 Budget Overview," Washington, D.C.: Air Force Financial Management and Comptroller, Deputy Assistant Secretary for Budget (SAF/FMB), March 2014; USCG, *Always Ready: United States Coast Guard 2013 Performance Highlights [and] 2015 Budget in Brief*, Washington, D.C., March 2014b; USCG, "Coast Guard Snapshot 2012," flyer, Washington, D.C., 2012a; U.S. Navy, "Status of the Navy," web page, August 26, 2015; William Lescher, "Navy FY15 Budget: Preserving Presence and Warfighting Capability," Navy Live blog, March 4, 2014; Ronald O'Rourke, "Navy Ford (CVN-78) Class Aircraft Carrier Program: Background and Issues for Congress," Washington, D.C.: Congressional Research Service, RS20643, October 22, 2013.

We addressed the first question by developing logic models to characterize the Coast Guard's missions, based on Coast Guard sources. To answer the second question, we collected and evaluated existing metrics, then mapped them to specific elements of the logic models. This enabled evaluation of the metrics in terms of how well they measure the elements and of how consistently and readily these measurements can be made. Finally, we identified potential metrics that either complemented or improved on existing metrics and developed a framework for review of metrics' values.

Chapter Two explains the above methodology in much greater detail. Chapters Three, Four, and Five describe the logic models, existing metrics, and potential metrics for each of the Coast Guard's 11 statutory missions. These missions are grouped under three headings: safety (Chapter Three), security (Chapter Four), and stewardship (Chapter Five). The section for each mission cites the Coast Guard sources that shaped our analysis of that mission. In Chapter Six, we review overarching themes and issues that span multiple missions. Chapter Seven presents a framework for reviewing the values associated with metrics in a structured, transparent way. Chapter Eight provides a brief recapitulation and some directions for future work. The 11 appendixes (one for each statutory mission) present the logic models, tables of existing metrics, and tables of potential metrics.

CHAPTER TWO

Methodology

Our research had four main components:

- **developing logic models** to ascertain what should be measured
- **analyzing existing metrics** and their relationships to those logic models
- **identifying potential metrics** that either refined or complemented existing metrics
- **composing a framework** for integrating and using metrics.

We describe these below.

Developing Logic Models

A basic prerequisite for evaluating and developing metrics is defining what they are intended to measure. This required comprehensive, structured logic models that described not only what the Coast Guard does but also how lower-level actions are mapped to higher-level goals. We developed them to be able to evaluate the extent to which individual metrics were capturing particular aspects of Coast Guard missions.

We structured our analysis around the 11 statutory missions, under three broad headings (based on Coast Guard Publication 3-0[1]):

- safety
  - search and rescue (SAR)
  - marine safety
- security
  - ports, waterways, and coastal security (PWCS)
  - drug interdiction
  - migrant interdiction
  - defense readiness
- stewardship
  - ice operations
  - aids to navigation (ATON) and waterway management (WWM)
  - marine environmental protection (MEP)
  - living marine resources (LMR)
  - other law enforcement (OLE).

---

[1] Coast Guard Publication 3-0, *Operations*, Washington, D.C.: U.S. Coast Guard, February 2012.

We recognize that there are other approaches for characterizing the Coast Guard's purview—for example, in terms of preparedness and response—but focused on the statutory missions for two principal reasons. First of all, the statutory missions are relatively fixed, while other taxonomies for characterizing the Coast Guard have been subject to more frequent changes. Second, other taxonomies are often more programmatic rather than mission-centric.

We began by developing a structure for the logic models that would characterize each statutory mission, drawing on prior RAND work and the literature on logic models.[2] We initially considered a simple structure involving inputs, outputs, and outcomes but then recognized the need to differentiate more finely among different types of inputs and outputs. The result is shown in Figure 2.1. *Strategic outcomes* relate to the three categories of missions mentioned above, namely safety, security, and stewardship. These may be closely aligned with the Coast Guard's overarching priorities, such as those enumerated in the *Coast Guard Western Hemisphere Strategy*,[3] *Arctic Strategy*,[4] and others.[5] Meeting strategic outcomes requires the fulfillment of mission-specific outcomes. *Outcomes* represent the culmination of a mission, such as lives saved or reduced risk from terrorism. Outcomes encompass more specific aspects of a mission, termed *accomplishments*. Fulfilling accomplishments requires *activities*, such as patrols or inspections. These are conducted with *assets*, such as helicopters; small boats; command, control, communications, computers, intelligence, surveillance, and reconnaissance (ISR) networks; units of personnel; and documentation. Those assets ultimately draw on *resources*, the Coast Guard's raw materials, such as recruits, funding, and legal authority.

Naturally, many external factors influence results at multiple levels of the logic model.[6] For example, for a given level of activity, economic and environmental conditions heavily influence the degree to which particular accomplishments, outcomes, or strategic outcomes can be achieved. Likewise, the Coast Guard has little ability to influence drug demand within the United States, living conditions in nearby other nations, or the number of individuals who become violent extremists. However, these will be powerful drivers of outcomes with respect to several Coast Guard missions and to its overarching priorities of securing borders and combating networks (as outlined in the *Coast Guard Western Hemisphere Strategy*[7]).

---

[2] See Victoria A. Greenfield, Valerie L. Williams, and Elisa Eiseman, *Using Logic Models for Strategic Planning and Evaluation: Application to the National Center for Injury Prevention and Control*, Santa Monica, Calif.: RAND Corporation, TR–370-NCIPC, 2006; Stephanie Young, Henry H. Willis, Melinda Moore, and Jeffrey Engstrom, *Measuring Cooperative Biological Engagement Program (CBEP) Performance: Capacities, Capabilities, and Sustainability Enablers for Biorisk Management and Biosurveillance*, Santa Monica, Calif.: RAND Corporation, RR-660-OSD, 2014; John A. McLaughlin and G. B. Jordan, "Logic Models: A Tool for Telling Your Program's Performance Story," *Evaluation and Program Planning*, Vol. 22, No. 1, 1999; and John A. McLaughlin, and G. B. Jordan, "Using Logic Models," in Joseph S. Wholey, Harry P. Hatry, and Kathryn E. Newcomer, eds., *Handbook of Practical Program Evaluation*, San Francisco: Jossey-Bass, 2004.

[3] USCG, *Western Hemisphere Strategy*, Washington, D.C., September 2014d.

[4] USCG, *Arctic Strategy*, Washington, D.C., May 2013c.

[5] Strategic outcomes could, for instance, be linked with the *Western Hemisphere Strategy*'s three priorities: combating networks, securing borders, and safeguarding commerce (USCG, 2014d). They could also be linked with the goals of USCG, *Command, Control, Communication, Computers, and Information Technology (C4&IT) Strategic Plan FY13–17*, November 2012b. On a geographically specific basis, strategic outcomes could be linked to *Arctic Strategy*'s strategic objectives, namely improving awareness, modernizing governance, and broadening partnerships (USCG, 2013c).

[6] Factors outside the context of the model are formally termed *exogenous* to it.

[7] USCG, 2014d.

Figure 2.1
Structure of the Logic Models

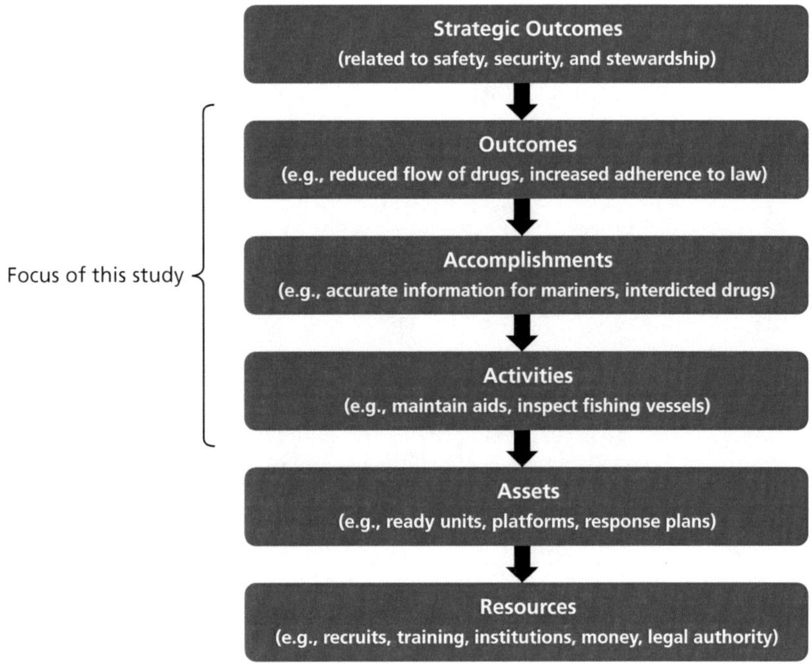

RAND *RR1173-2.1*

Given that this report is intended to analyze operational-level metrics, we developed the three levels of the logic models that are relevant to operations: activities, accomplishments, and outcomes. For brevity, we collectively refer to these as the *elements* of the logic models.

We developed each of the 11 logic models (one for each statutory mission) by researching numerous Coast Guard documents and interviewing dozens of subject-matter experts (SMEs).[8] Chapters Three, Four, and Five describe the primary documents we used for each mission and the reasons we developed individual logic models as we did. Our interviews included discussions with personnel from Coast Guard Headquarters, Atlantic Area, Pacific Area, all nine districts, Force Readiness Command, and the Intelligence Center. Moreover, we provided drafts of our logic models to stakeholders twice, enabling us to use their feedback to iteratively refine the logic models. The Coast Guard has *not*, however, doctrinally approved the logic models. Chapters Three, Four, and Five also present the logic models, which are reprised in the appendixes for convenient cross-referencing with the metrics.

The logic models are intended to characterize the current state of Coast Guard outcomes, accomplishments, and activities. Naturally, these may be subject to change due to shifts in national, Department of Homeland Security (DHS), or Coast Guard goals or priorities, as well as emerging threats or technological changes. As the Coast Guard evolves, the logic models will likewise need to be modified periodically.

---

[8] One challenge we encountered is that the elements of different missions are not perfectly disjoint; for example, a single patrol may deter multiple types of illegal activity. However, the vast majority of mission elements were disjoint enough to enable separate analysis of each mission.

Analyzing Existing Metrics

While developing the logic models, we also drew on many of the same documents and experts to ascertain what metrics the Coast Guard was currently using. Ultimately, we found 157 widely shared metrics spanning all 11 missions. We recognize that other metrics are being recorded within specific commands; experts mentioned some of these to us, and the list of potential metrics that we later identified incorporated these. However, in analyzing existing metrics, we are counting only metrics that are being communicated beyond the confines of a small community. The 157 metrics that we included are all being widely disseminated; we did not analyze the extent to which these are being used for decisionmaking. Incidentally, some may be used for purposes other than measuring operational performance. We did not evaluate their utility in other contexts.

Next, we mapped each of the existing metrics to the outcomes, accomplishments, and activities within the logic models. This enabled us to determine which elements of the logic models were currently being measured and which were not. In addition, it enabled us to evaluate how well the elements were being measured. As in prior RAND work,[9] we applied three criteria to each metric:

- **validity**—the extent to which the metric captures the outcome, accomplishment, or activity being assessed
- **reliability**—how consistently measurements can be made
- **feasibility**—how easily the measurement can be made.[10]

We set up a simple three-point scale, with high, medium, and low values for each metric. The details of this scale are shown in Table 2.1.

Chapters Three, Four, and Five describe the existing metrics. The appendixes also list these and provide their ratings.

Table 2.1
Criteria for Evaluation of Metrics on a Three-Point Scale

	Validity	Reliability	Feasibility
High	The metric directly measures the element	Quantitative, well defined, and stable	Required data sets are available and well organized
Medium	The metric is closely related to the element[a]	Either qualitative, well defined, and stable, or quantitative, less well defined, and somewhat volatile	Required data sets could be collected without much difficulty
Low	The metric is indirectly related to the element	Qualitative, dependent on judgment, anecdotal	The required data sets would be challenging to collect

[a] This may include proxies that are closely correlated with the element.

---

[9] See Young et al., 2014.

[10] An often-overlooked aspect of feasibility is that measurements requiring manual data entry are susceptible to error, particularly if the person required to enter the data sees this responsibility as a distraction from operational tasking (which may make them more careless). Several of our interlocutors expressed concern about the quality of data in Coast Guard databases due to busy operators being focused on other tasks.

Identifying Potential Metrics

After evaluating the existing metrics, we developed sets of potential metrics for two main purposes. We intended most of these metrics to measure elements of the logic models that are not currently being measured. (Many of these center on the less-visible parts of missions, for example, those relating to interagency or international partnerships.) These potential metrics can complement existing metrics to characterize operational performance more comprehensively.

The remaining potential metrics are intended to improve on existing metrics in some way or to provide further nuances. Many of these represent slight modifications of existing metrics. For example, some leverage existing metrics that measured absolute numbers of items but use an appropriate denominator to generate a percentage or a rate. Others include qualifying phrases that adjust the scope of a metric, while still others take a single metric and divide it into multiple closely related metrics. A few involve variations on a theme: Where an existing metric measures one item, such as a quantity of cocaine, the potential metrics count similar items, such as quantities of other drugs.

As with other parts of this work, we drew on a combination of documents, discussions with experts, and our team's expertise to develop potential metrics. Some of the experts we consulted mentioned metrics they were using within their districts but that were not currently promulgated throughout the Coast Guard; we included some of these as potential metrics.

We mapped all the potential metrics to elements of the logic models where such relationships were not already clear. We also evaluated the potential metrics in terms of validity, feasibility, and reliability, just as we had done with the existing metrics. In some cases, this evaluation contributed to the refinement or replacement of potential metrics that did not score well. However, we retained metrics with low scores when we did not perceive a clearly higher-scoring way to measure a given element or when metrics with low scores could complement others to compensate for their deficiencies.

Chapters Three, Four, and Five describe the potential metrics. The metrics are also listed in the appendixes, along with their ratings.

Composing a Framework

Finally, we developed a framework for reviewing the values associated with metrics in a structured fashion. Since the methodology associated with developing this framework is deeply interwoven with the description of the framework itself, we will describe that methodology in Chapter Seven amid discussion of the framework as a whole.

CHAPTER THREE

Safety Missions

This chapter discusses the logic models, existing metrics, and potential metrics associated with two missions that center on safety: search and rescue (SAR) and marine safety.[1] For each, we first describe the mission in a few lines. Next, we present the logic model and explain how it was developed, based on Coast Guard sources. After this, we highlight aspects of the existing metrics that are of particular interest, capturing common themes and issues. Finally, we briefly describe the potential metrics we identified for each mission. Some of them represent refinements of the existing metrics, and we explain how they have been modified. Others measure elements of the logic model that had not previously had metrics associated with them; we describe the relationships of these metrics to the elements they measure, as well as any opportunities or limitations associated with them. This chapter does not include comprehensive lists of the metrics themselves, which are very long; complete lists of existing and potential metrics, as well as evaluations and comments, appear in the appendix associated with each mission.

Search and Rescue

The purpose of the SAR mission is to respond to maritime incidents that have the potential to cause deaths, injuries, and property damage or loss and thereby to mitigate the effects of such incidents.[2]

Development of the Logic Model

In developing the logic model for SAR (Table 3.1), the primary source document was Commandant Instruction (COMDTINST) M16130.2F.[3] This instruction clearly outlines the current metrics and the associated organizational targets to measure success. Additional source documents, such as the Coast Guard's fiscal year (FY) 2015 Strategic Planning Direction (SPD) and the Atlantic Area FY 2014 Operational Planning Direction (OPD),[4] all align with

1 Coast Guard Publication 3-0, 2012.

2 Efforts to prevent property damage or loss may also be combined with trying to minimize the environmental impact of the incident. Cleanup of a large-scale incident, however, is part of the MEP mission.

3 Commandant Instruction Manual 16130.2F, "The United States Coast Guard Addendum to the United States National Search and Rescue Supplement (NSS) to the International Aeronautical and Maritime Search and Rescue Manual (IAMSAR)," Washington, D.C.: U.S. Coast Guard, January 7, 2013.

4 Charles D. Michel, "Fiscal Year 2015 Strategic Planning Direction," memorandum, August 13, 2014; R. C. Parker, "FY14 Operational Planning Direction," memorandum, Portsmouth, Va.: U.S. Coast Guard Atlantic Area, November 27, 2013.

Table 3.1
Logic Model for Search and Rescue

Outcomes	Accomplishments	Activities
1. Save and assist lives and property	1.1. Establish communication channels and monitor communications lines	1.1.1. Place and employ communications infrastructure to monitor maritime environment
		1.1.2. Educate maritime community on available resources and how to employ them
		1.1.3. Notify search-and-rescue unit (SRU) in a timely manner
	1.2. Maintain ready assets and crew for response	1.2.1. Posture assets (e.g., helicopters, boats, cutters) to respond effectively and rapidly
		1.2.2. Ensure that responding crews are able to coordinate with other responding units and agencies when on scene
	1.3. Execute search response operations	1.3.1. Develop search action plan and conduct operational risk evaluation, based on available information and appropriate software
		1.3.2. Deploy the appropriate SRU to execute the response
	1.4. Improve coordination with and capabilities of non–Coast Guard SAR entities	1.4.1. Develop SAR memorandums of understanding (MOUs) with strategic partners
		1.4.2. Foster bilateral, regional, interagency, international, private-sector, and other external SAR engagement
		1.4.3. Assist in the development of SAR policy, requirements, programs, and plans
		1.4.4. Establish domestic and international partnerships to save lives in the most effective manner
		1.4.5. Coordinate Automatic Maritime Vessel Emergency Response (AMVER)
		1.4.6. Respond effectively to surface picture (SURPIC) requests

COMDTINST M16130.2F without significant deviations. The Coast Guard's SAR Program outlines four key objectives that inform and shape the activities, accomplishments, and outcomes:

- minimize loss of life, injury, and property loss and damage in the maritime environment
- minimize crew risk during SAR missions
- optimize use of resources in conducting SAR
- maintain a world leadership position in maritime SAR.

To effectively evaluate success against the backdrop of these objectives, the Coast Guard uses four primary "performance metrics" or "outcome metrics" as promulgated in guidance and policy:

- percentage of people in imminent danger saved in the maritime environment
- percentage of all people in distress saved after Coast Guard notification

- percentage of property "in danger of loss" saved
- property (in millions of dollars) at risk.

At an early stage of our research, we considered having two distinct outcomes: one each for saving lives and saving property. We eventually decided to combine them into a single "save and assist lives and property" outcome. This was based on the fact that activation of the SAR response system is identical for rescuing both people and property. (Naturally, lifesaving is *always* the priority when choices must be made between the two.) The precise response actions involved will be influenced by the nature of the distress, but the broadly defined outcome, accomplishments, and activities associated with response are the same.

Existing Metrics

Currently, the Coast Guard divides the number of lives and value property saved by the total number of lives and value of property at risk to measure mission effectiveness. In doing so, it focuses exclusively on situations in which ten or fewer lives, and less than $2 million in property, were at risk. Performance targets are based on historical performance and reasonable expectations of success. The Coast Guard recognizes that the SAR environment is not static, and that performance targets should increase as technology develops and improvements are implemented in the SAR system. In this context, the Coast Guard systematically evaluates and updates the SAR performance expectations with periodic adjustments.

We evaluated existing metrics based on validity, reliability, and feasibility in measuring SAR performance. The majority of the metrics were assessed to be of high validity (the metric accurately reflects what is intended to be measured). Reliability (consistent application of a standard) was assessed as medium for most metrics because such terms as "imminent danger," "distress," and "in danger of loss" may be subject to a certain degree of interpretation (however, a ten-factor flowchart in COMDTINST M16130.2F and the headquarters-maintained property calculator program help standardize assessments).

In addition to these metrics, certain other metrics were characterized as "General SAR Program Standards and Requirements" in source documents, including the following:

- percentage of continuous coverage of reception of a 1-W signal from a 1-m antenna, out to 20 nmi from shore, as part of the National Distress and Response System
- percentage of cases in which "suitable" SAR resources were ready to proceed within 30 minutes of notification of distress
- percentage of cases in which the Coast Guard SRU arrived on scene within two hours of notification.

Validity, reliability, and feasibility for each of these were evaluated as either high or medium, with one exception: The metric for effective coverage of reception of a 1-W signal from a 1-m antenna was deemed to have low feasibility.

Potential Metrics

Overall, we feel that the current metrics provide the elements necessary to assess basic performance of the service's ability to rescue mariners and equipment in distress. The proposed additional metrics are offered as a means of making a more-comprehensive assessment of SAR mission performance and of evaluating standards that are already required or implied but not yet measured. Potential metrics to help communicate other important aspects of the SAR mission, especially as these relate to efficient use of resources and crew risk, include the following:

- the percentage of cases in which an incident's response requirements exceeded organic Coast Guard capabilities
- the percentage of cases in which actions (such as notifying a response unit or collecting more information) are taken within five minutes
- the percentage of cases in which the SAR object (person or vessel) is located within the search area
- the percentage of asset-hours and personnel-hours spent responding to false alerts and hoaxes.

In addition, as the logic model demonstrated, partnerships are an important part of the SAR mission. We therefore introduced several metrics relating to collaboration with other U.S., international, and private-sector partners.

As noted earlier, existing SAR metrics captured only smaller-scale incidents, involving ten or fewer people and less than $2 million in property. There are good reasons for focusing on smaller-scale incidents, since a single large-scale one could skew average values in a way that would obfuscate the actual state of affairs. However, we proposed adding separate metrics for larger-scale events, so that such events and the trends related to them would not be missed altogether. The property threshold could also be indexed to inflation (and perhaps revised upward to reflect the value of $2 million at an earlier date).

Likewise, we found that some existing metrics could be revised to reflect the circumstances of particular districts. For example, the vast distances within Districts 14 and 17 make it impossible for Coast Guard SAR assets to reach particular areas within the stated two-hour time frame. Applying the existing metric (for arrival within two hours) to remote parts of these districts does not reflect their geographic dispersion. Instead, we have restated the metric as, "percentage of cases in which Coast Guard SRU arrives on scene within two hours of notification *or within a logistically feasible time limit*."[5]

Marine Safety

The focus of the Coast Guard's marine safety mission is on the prevention of deaths, injuries, property loss, and environmental damage from activity on or near the water.[6]

Development of the Logic Model

In developing the logic model (shown in Table 3.2), we drew heavily on the FY 2015 SPD, area OPDs, the Marine Safety and Prevention Performance Plan, and the Coast Guard Busi-

[5] Some SMEs had reservations about whether relaxing the two-hour constraint in specific geographic cases could lead to its relaxation elsewhere, with dangerous consequences. They also indicated that showing a lack of ability to fulfill the two-hour limit was important for the awareness of Congress and other external stakeholders. However, as an internal operational measure, applying a two-hour constraint where it cannot be met would seem to distort perceptions.

[6] One of our Coast Guard interlocutors noted that some accomplishments and activities that appear to be outside the statutory missions, such as response to a major vessel grounding that does *not* lead to a spill, could conceivably be included under the marine safety rubric. However, current documentation does not categorize this as such.

Table 3.2
Logic Model for Marine Safety

Outcomes	Accomplishments	Activities
1. Reduce the frequency and extent of injuries, fatalities, property damage, and environmental damage due to safety incidents	1.1. Increase compliance by mariners, marine facilities, and vessels	1.1.1. Monitor, inspect, examine, and investigate facilities, vessels, and contents (such as containers, bulk goods, and ballast water)
		1.1.2. Cite non–incident-related safety work list items from facility and vessel inspections and note them accordingly in the Maritime Information for Safety and Law Enforcement (MISLE) casework narrative
	1.2. Increase the stakeholder knowledge base and its documentation	1.2.1. Provide consultation to industry advisory bodies and conduct maritime sector outreach
		1.2.2. Administer a licensing and credentialing program for professional mariners and vessel documentation and decal programs
		1.2.3. Provide regulatory, safety, and compliance guidance, as well as information on best practices, to all stakeholders
		1.2.4. Monitor and report North Atlantic iceberg conditions using fixed-wing aircraft and reports from ships as part of the International Ice Patrol
2. Investigate and document safety incidents	2.1. Investigate incidents	2.1.1. Document incident and collect evidence
		2.1.2. Determine incident causal factors
		2.1.3. Identify responsible parties
		2.1.4. Assist interagency, state, and local investigations
		2.1.5. Disseminate investigation findings for use in improved regulation to field and industry
	2.2. Prepare for legal proceedings	2.2.1. Issue appropriate citations
		2.2.2. Prepare case packages for prosecution

ness Intelligence (CGBI) Operational Planning Assessment Report (OPAR).[7] We used these to identify two broad outcomes: reducing the frequency and extent of safety incidents and investigating safety incidents that have occurred. Each of these entails two accomplishments. Increasing compliance (by observing and citing potential risks) and increasing stakeholder knowledge contribute to the first outcome.[8] For the second outcome, the two accomplishments are investigating the incident—through such activities as identification of responsible parties and causal factors—and preparing for legal proceedings by issuing citations and preparing case packages.

[7] Michel, 2014; W. D. Lee, "FY2015 LANTAREA Operational Planning Direction," memorandum, Portsmouth,Va.: U.S. Coast Guard Atlantic Area, September 19, 2014. The OPAR is a quarterly summary of data from the CGBI.

[8] We recognize that there may be instances in which compliance with specific laws or regulations does not actually reduce the frequency or extent of negative events. Evaluation of the impact of laws or regulations is a separate issue.

Perhaps surprisingly, we were informed that the Coast Guard's participation in the International Ice Patrol—monitoring icebergs in the North Atlantic, a responsibility engendered by the *Titanic* disaster—is part of the marine safety mission. We included it under the rubric of increasing stakeholder knowledge, since the Coast Guard is ultimately providing information to improve safety.

Existing Metrics

The existing metrics we identified had a very narrow focus; they collectively addressed only two of the 18 elements of the logic model. Nine of them related to outcome 1 (reducing the frequency and extent of incident consequences, including fatalities, injuries, and spills). The metrics relating to human casualties scored highly in terms of validity, reliability, and feasibility. However, the metrics relating to spills scored medium in terms of validity and reliability. The validity rating reflects the fact that counting the number of spills, without reference to magnitude except for measuring the fact that they exceed a very low threshold, provides limited information on overall consequences. Moreover, nonoil chemical spills are counted collectively, without taking into account that some chemical agents are much more harmful than others. There may also be some limitations on the accuracy with which numbers of spills can be characterized because of uncertainty about whether a particular spill exceeded a given threshold or because of incomplete dissemination of information. While fatalities are overwhelmingly likely to be reported, spills may be undercounted to varying degrees.

The other 26 metrics all related to activity 1.1.1 (monitoring, inspection, examination, and investigation of facilities and vessels). Most of these metrics focus on the number of safety inspections, examinations, and monitoring activities performed on vessels and marine facilities. All these metrics have high reliability and feasibility; they are easy to count consistently. However, since each addresses only one portion of the activity, the validity is typically medium. Collectively, they provide a lot of information about the activity, but none of the individual measurements wholly captures it.

Potential Metrics

Many of the potential metrics we developed draw on the same data sets used in existing metrics, but using rates or percentages as opposed to absolute numbers, and more finely parsing the data. This introduces some uncertainty, since denominators (such as the number of passenger trips taken) may not be perfectly known, but helps to avoid misperceiving rises in traffic as rises in accident rates.

As indicated above, existing metrics do not capture 16 out of the 18 elements of the logic model. Most of the potential metrics therefore aim to measure the missing elements of the logic model, such as outreach and investigations. Some of these have medium or low feasibility, since data sets would need to be gathered through surveys, modeling, or knowledge testing. However, others are high in terms of all three of our criteria, such as the percentage of cases in which causal factors are determined following an incident.

Developing metrics for activities 1.2.1 (provide consultation and conduct outreach) and 1.2.3 (provide guidance and information to stakeholders) was challenging. This was a recurring problem across multiple missions; it is inherently difficult to measure communication, collaboration, partnerships, and other aspects of interaction. We addressed this by developing a complementary mix of metrics. Some of these had high reliability and feasibility but limited validity, such as forms of communication being used or numbers of meetings. Others, such as estimated percentages of audiences being reached, would require surveys, limiting their feasibility; however, they had higher validity. Alternatively, estimates could be generated that were not based on surveys, but with a lower degree of reliability. For other missions in which partnerships and communication played a more central role, we developed more numerous and varied metrics relating to these items, as will be discussed in the context of the missions.

Some of these metrics could be further specified in terms of the context in which incidents took place. For example, as some of our interlocutors noted, recreational boating injuries in small, isolated bodies of water could be grouped separately from those occurring in more densely trafficked environments in which the Coast Guard has an active presence.

Summary of Findings on Safety Missions

Aggregating across the two safety missions (SAR and marine safety), we found that existing metrics do not capture most of the elements of the logic model (Table 3.3). This is not surprising, given that the metrics predate the logic model.

All but one of the existing metrics for these missions scored "high" in terms of feasibility, meaning that they were relatively easy to measure. However, they had more limitations in terms of the other criteria we used. Only 14 percent of the marine safety metrics had high validity (though 86 percent of the SAR metrics did). While nearly all of the Marine Safety metrics had high reliability, this was the case for only one-third of the SAR metrics.

Many of the safety missions' potential metrics complement the existing metrics by capturing elements of the logic models that existing metrics do not measure. Others address opportunities to refine some of the existing metrics, while leaving others in place as they are (as outlined in the individual sections on SAR and marine safety).

Table 3.3
Numbers of Safety Mission Logic Model Elements and Percentages Measured by at Least One Metric

Type of Element	Number	Percentage
Outcomes	3	67
Accomplishments	8	0
Activities	26	12
Total	37	14

CHAPTER FOUR

Security Missions

This chapter covers the logic models, existing metrics, and potential metrics associated with four missions that fall under the broad rubric of security: PWCS, drug interdiction, migrant interdiction, and defense readiness.[1] As in the previous chapter, we begin by very briefly describing each mission, then present its logic model and an explanation of how it was developed. We then broadly describe the existing metrics for each mission and their evaluations according to our three criteria: validity, reliability, and feasibility. Finally, we describe key aspects of the potential metrics, including their evaluations. As noted in the previous chapter, the complete tables of both existing and potential metrics appear in the appendixes, rather than in the body of the text.

Ports, Waterways, and Coastal Security

The Coast Guard's PWCS mission is intended to develop robust maritime security regimes, achieve maritime domain awareness, and conduct maritime security and response operations (MSROs) to detect, deter, and disrupt maritime terrorist attacks, respond to and recover from attacks that may occur, and work with port partners to review vessel and security plans.

Development of the Logic Model

Our development of the logic model for the PWCS mission (shown in Table 4.1) relied on a broad selection of documents. This reflects the fact that the PWCS mission itself encompasses a wide variety of tasks, some as routine as ensuring that port workers are properly identified and some as demanding as responding to a terrorist attack on the Statue of Liberty. The *Maritime Prevention Program Performance Plan* details the Coast Guard's Port Security program, intended to improve performance for the various activities and accomplishments that comprise PWCS.[2] This includes DHS's *Small Vessel Security Strategy*,[3] the Port Security engagement strategy, improvements in cargo security, and other courses of action. Another valuable source was the MSRO Scorecard,[4] whose comprehensive list of activities gave an illuminating picture

[1] Coast Guard Publication 3-0, 2012.

[2] Joseph Servidio, *U.S. Coast Guard Maritime Prevention Program Performance Plan, Fiscal Years 2014–2019*, Washington, D.C.: U.S. Coast Guard, August 2013b, Not available to the general public.

[3] U.S. Department of Homeland Security, *Small Vessel Security Strategy*, Washington, D.C., April 2008.

[4] Commandant Instruction Manual 16600.6, Maritime Security Response Operations Manual, Not available to the general public.

Table 4.1
Logic Model for Ports, Waterways, and Coastal Security

Outcomes	Accomplishments	Activities
1. Enhance preparedness for and prevention of maritime terrorist attacks	1.1. Ensure vessel and facility compliance with maritime security laws and regulations	1.1.1. Ensure plans are in place and review to ensure adequacy
		1.1.2. Conduct inspections and boardings to ensure plans are followed
		1.1.3. Enforce transportation worker identification credentialing (TWIC)
		1.1.4. Conduct external patrols to assess security measures
	1.2. Collect intelligence in collaboration with other agencies and entities	1.2.1. Collect intelligence
		1.2.2. Contribute to collaborative bodies and organizations to analyze, integrate, and disseminate intelligence
		1.2.3. Share relevant intelligence information with other stakeholders, as appropriate
	1.3. Ensure PWCS preparedness	1.3.1. Conduct Coast Guard exercises
		1.3.2. Conduct exercises with external entities
		1.3.3. Conduct outreach with the public
		1.3.4. Engage with the other members of the Area Maritime Security Committee (AMSC)
2. Deter potential maritime terrorist attacks; counter and respond to actual maritime terrorist attacks	2.1. Maintain presence near critical or vulnerable targets	2.1.1. Conduct waterborne, air, or shoreside patrols around high-risk infrastructure or events
		2.1.2. Escort vessels
		2.1.3. Conduct random and targeted security boardings of vessels
		2.1.4. Enforce fixed security zones
	2.2. Counter terrorist attacks when cued by intelligence or events	2.2.1. Surge to respond to security threats when alerted
		2.2.2. Provide transit protection support (TPS)

of the diverse requirements the PWCS mission entails. Vessel escorts (including nuclear submarine escorts), boardings, positive control measures on high-interest vessels and vessels carrying certain dangerous cargoes, and patrols around security zones and possible infrastructure targets are only a few of the activities crucial to the PWCS mission.

We identified two main desired PWCS outcomes: enhancing preparedness for and prevention of maritime terrorist attacks and deterring, countering, and responding to potential or actual maritime terrorist attacks. Initially, we considered the possibility that deterrence and countering and responding should be two different outcomes but ultimately decided that actions taken in pursuit of countering an attack are inherently part of deterring that attack.

Although some of our source documents identified the development of security regimes and partnerships with ports as strategic goals of the PWCS mission, we moved these initiatives to the accomplishment level because they support the larger goals of preparedness and response. Other accomplishments concern themselves with intelligence collection, security exercises, maintaining presence and/or protection near potential targets, and countering terrorist attacks when cued by intelligence or events. Activities associated with these accomplishments were, in some cases, drawn from the MSRO Scorecard and other documents that detail the metrics the Coast Guard currently uses to assess its performance. In other cases, interviews with personnel at the district and headquarters levels were extremely useful for identifying activities, given scant mention in the documentation. These conversations were particularly useful, for example, in ascertaining the intelligence-related activities currently supporting the PWCS mission.

Existing Metrics

Existing PWCS metrics concentrate around ensuring compliance with port facility security laws and regulations and are almost uniformly rated as high across the spectrum of validity, reliability, and feasibility. One exception is the metric of maritime security risk reduction with respect to consequence management, terrorist transfer, or weapons of mass destruction transfer, which is rated as low in feasibility. It would be extremely difficult to estimate the extent to which Coast Guard actions result in some percentage of reduced risk in this sense. Not only are such events (thankfully) rare enough that any metric relying on them will be inherently unreliable, but Coast Guard performance would also rely on the actions of any number of other groups, from local law enforcement to DoD intelligence analysts.

Additionally, performance metrics related to Maritime Force Protection Unit Compliance with the USN escort requirements are maintained separately from the MSRO Scorecard. These metrics measure the Coast Guard's ability to provide the required force package for the USN ballistic missile submarine in-transit escort.

On the whole, while there are some good activity-level PWCS metrics, there are issues relating to metrics at the accomplishment and outcome levels. Estimates of risk reduction are based on SME evaluations of aggregated scenarios, limiting reliability. Also, while existing metrics cover certain elements of the PWCS logic model, there are gaps in others that could be covered by complementary metrics.

Potential Metrics

Our potential metrics address such areas of the logic model as intelligence collection and sharing, outreach to security partners, transit protection support, and surge activity in the event of a crisis. In creating the new series of metrics, we drew heavily from our source documents, particularly the FY 2015 SPD,[5] which emphasizes deterrence, interdiction, and small-vessel security.

Metrics for the PWCS outcome of enhancing preparedness for and prevention of maritime terrorist attacks are difficult to design. We have proposed an estimated risk reduction associated with intelligence preparation and enhanced compliance with security regulations, although it rates low on reliability and feasibility because of the complexity of calculating such a percentage. Risk modeling would enable at least a basic calculation, however.

The potential metrics for accomplishments and activities associated with this outcome rate more highly across the spectrum; most are medium or high. Many new metrics are based on percentages with clear and knowable denominators, such as the percentage of permanent

[5] Michel, 2014.

and nonpermanent marine critical infrastructure plans reviewed and, of those, the percentage that comply with security requirements and regulations. A few potential metrics would require surveying external audiences for Coast Guard–generated intelligence to determine their opinion of the product. Surveys can be challenging to administer successfully; this fact is reflected in the low feasibility of metrics that require them.

Currently, no metrics exist for gauging the effectiveness of exercises conducted either by the Coast Guard alone or jointly with other agencies. We propose several, most of which rate highly on reliability and feasibility and medium on validity; most are counts of exercises performed singly or with partners. These counts could very well be different for each district, which will likely have a number of factors influencing how, when, how often, and with whom it runs exercises. One proposed metric, the estimated risk reduction due to exercises and outreach, rates low on reliability and feasibility because of the difficulty of accurately making such a measurement.

We have suggested several new metrics for the outcome of deterring, countering, and responding to potential maritime terrorist attacks. Three of these—the number of injuries, number of fatalities, and extent of property damage associated with a maritime terrorist attack—are rated as low on validity because these outcomes depend on many factors beyond Coast Guard control, not least the intent, resources, and capabilities of terrorists. However, these metrics rate highly in terms of reliability and feasibility.

We have also proposed several metrics for accomplishments and activities associated with this outcome. Roughly one-half have low reliability or feasibility, mainly because they will require wargaming or exercises or because they require ranking potential targets in ways that may be difficult or not ultimately feasible. (They could also be used to evaluate wargames or exercises that were being conducted for other purposes.) Our potential metric of calculating percentages of time that the Coast Guard is present near critical or vulnerable targets or that high-risk infrastructure or events are patrolled, for example, would result either in rating all potential targets as equally vulnerable or in ranking potential targets according to risk (as the Coast Guard already does to some extent). Our remaining potential metrics, however, rank medium or high on all counts. Again, we turn to percentages with clear and knowable denominators to determine the level of support that the Coast Guard provides, whether requested of it (in the case of vessel escorts) or not (in the case of response to maritime terrorist attacks).

Drug Interdiction

The Coast Guard's drug interdiction mission "supports national and international strategies to deter and disrupt the market for illegal drugs, dismantle Transnational Criminal Organizations (TCOs)/Drug-Trafficking Organizations (DTOs), and prevent transnational threats from reaching the U.S."[6] Moreover, the Coast Guard is the lead federal agency for combating maritime drug trafficking. It is a key contributor to Joint Interagency Task Force–South (JIATF-S), whose primary purpose is to conduct detection and monitoring to counter illicit trafficking. The Coast Guard's contributions include both patrol assets and staff. JIATF-S is generally led by a Coast Guard flag officer, and is subordinate to the U.S. Southern Command (USSOUTHCOM).

---

[6] USCG, 2014b, p. 11.

Developing the Logic Model

In developing the logic model for drug interdiction (see Table 4.2), our primary source document was the *Law Enforcement Operations Program Performance Plan*.[7] This document outlines the current metrics and the associated organizational targets to measure success. Additional source documents, such as the 2015 *Coast Guard Strategic Planning Direction* and 2015 *Atlantic Area Operational Planning Direction*, are well aligned with it.[8]

Using these documents, we developed two outcomes. The first is to reduce the maritime flow of illegal drugs. In addition to the obvious direct effect of curtailing drug supply, this also helps to counter drug-trafficking and TCOs both by imposing costs on them and by apprehending their personnel (who may also provide follow-on intelligence). The best-known activities for reducing the maritime flow of drugs—detecting, monitoring, intercepting, interdicting,[9]

Table 4.2
Logic Model for Drug Interdiction

Outcomes	Accomplishments	Activities
1. Reduce the maritime flow of illegal drugs	1.1. Deter, divert, detain, and disrupt maritime drug-smuggling flows	1.1.1. Detect and monitor
		1.1.2. Intercept
		1.1.3. Interdict
		1.1.4. Board and apprehend
		1.1.5. Achieve visible or perceived presence
	1.2. Support and coordinate with other agencies and international partners to counter maritime drug flows (achieving unity of effort)	1.2.1. Conduct joint operations (including shiprider operations) and exercises with other agencies and nations
		1.2.2. Assist partner nations in the development of capabilities to counter drug flows
		1.2.3. Develop MOUs and memorandums of agreement (MOAs) with other agencies and nations to enable more-effective counterdrug operations
	1.3. Enable prosecution of smugglers	1.3.1. Collect and handle evidence for prosecution purposes
		1.3.2 Document case details and prepare case packages for prosecution
2. Increase intelligence and situational awareness of maritime drug flow	2.1. Conduct ISR activities to counter drug flows and trafficking networks	2.1.1. Investigate vessel and other physical evidence
		2.1.2. Interview apprehendees
		2.1.3. Deploy ISR systems
		2.1.4. Collect and integrate ISR systems' data
		2.2.1. Engage in interagency bodies and liaise with other agencies to create actionable intelligence, enable information sharing, and foster intelligence collaboration

[7] USCG, *Maritime Law Enforcement Operations Program Performance Plan FY2014–2019*, August 2013e.

[8] Michel, 2014; Lee, 2014.

[9] Across much of the Coast Guard, *interception* means pursuing the target, while *interdiction* means stopping it. (These definitions may vary across communities and individuals.)

boarding, and detaining drug-laden vessels—are grouped under a single accomplishment. Under the same accomplishment, achieving visible or perceived presence can also help to divert drug flows to less-desirable pathways for the smugglers and cause some drug-laden vessels to have to turn around, at least temporarily.[10] Two other important accomplishments supporting this outcome are collaborating with partners and enabling prosecution of smugglers.

The other outcome focuses on collecting intelligence and increasing situational awareness. We included this as a distinct outcome to reflect the fact that drug smuggling involves complex networks that need to be countered through detailed ISR.

The significance of the less-tactical elements of the logic model is highlighted by the fact that they contribute to the priorities and objectives stated in high-level documents. For example, the strategic objectives of the Office of National Drug Control Policy's *National Southwest Border Counternarcotics Strategy* include disrupting trafficking organizations through prosecutions and enhancing cooperation with Mexican authorities.[11] Likewise, the Coast Guard's *Western Hemisphere Strategy* cites combating networks as one of its three top priorities, requiring intelligence, partnerships, and helping partners with capacity building. Another one of the top priorities—securing borders—includes ample discussion of awareness, partnerships, capacity building, and threat prioritization, as well as tactical interdiction.[12]

Existing Metrics

Currently, the Coast Guard measures both its ability to effectively interdict the flow of drugs and its ability to deploy required assets along known threat vectors. In accordance with national strategic guidance, the Coast Guard's efforts and measures focus primarily on its ability to combat the flow of cocaine. Along these lines, an existing metric that receives considerable attention is the quantity of cocaine interdicted.

We evaluated existing metrics in terms of their validity, reliability, and feasibility. The majority of the metrics were of high validity (the metric accurately reflects what is intended to be measured). Reliability (consistent application of a standard) was medium for most metrics. While counting days deployed and resource hours results in a high degree of reliability, there is less certainty with rates, percentages, and quantity of flow because of the "unknown denominator" factor: Estimates of the underlying flow rate of drugs routinely have varying degrees of accuracy, but the rate is always imperfectly known. Although the Consolidated Counterdrug Database (CCDB) provides a comprehensive and accepted method of estimating the flow of cocaine, there are inherent uncertainties. The feasibility of these metrics is high because numerous databases are available from which to extract the required information.

---

[10] Rerouting of drug flows can affect not only drug availability but also the security of states in the region (either positively or negatively). For example, effective enforcement at sea may drive drug smugglers to move the drugs via the Central American isthmus, exacerbating governance challenges there. However, the same actions may enhance the stability of Caribbean nations that would otherwise be more extensively used as way stations.

[11] Office of National Drug Control Policy, *National Southwest Border Counternarcotics Strategy*, 2013. The strategic objectives also explicitly include interdicting drugs in the maritime domain.

[12] USCG, 2014d.

Potential Metrics

The current metrics meet the service's foundational needs to assess performance as it relates to combating the flow of cocaine and asset availability. The following potential metrics offer means of assessing Drug Interdiction mission performance more comprehensively:

- quantity of drugs removed, by type (not just cocaine)
- seizures, such as weapons and cash
- number of cases in which the Coast Guard is unable to properly respond, detect, or monitor targets
- various metrics associated with evidence handling, case packaging, and successful prosecution by the Department of Justice (DoJ)
- number of events in which intelligence led Coast Guard assets to a drug vessel.

Many of these potential metrics can be extracted from information already being collected but not widely disseminated throughout the Coast Guard.

Several potential metrics would help facilitate sharing information with the various units and agencies that provide intelligence to responding assets. Several interviews suggested that the Coast Guard shares intelligence very well, but information also needs to flow in the other direction.

Several of the potential metrics expand the scope of illicit activity being measured beyond cocaine. The metrics associated with the movement of marijuana, methamphetamines, cash, and weapons are potential sources of leading and trailing indicators of the overall effectiveness of the entire drug interdiction enterprise. Drugs other than cocaine are also important sources of revenue for DTOs. Also, in some districts, these drugs constitute an important or dominant portion of the maritime drug flow. While cocaine interdiction remains a national goal, and one for which national and Coast Guard goals are set, data on other drugs can also be relevant.

Finally, tracking such interactions as training exercises and bilateral agreements can provide insights into the quality of partnerships, mutual contributions to collective capabilities, and information sharing in support of drug interdiction efforts.

Migrant Interdiction

The Coast Guard's migrant interdiction mission is intended to curtail the flow of undocumented migrants entering the United States via maritime routes.

Development of the Logic Model

The most important source for the logic model for the migrant interdiction mission (see Table 4.3) was the Coast Guard's *Migrant Interdiction Mission Performance Plan FY 2011–2016*.[13] In developing the logic model, we also drew on the SPDs from FY 2014 and 2015, along with OPDs for FY 2014 from the Atlantic Area and District 7.[14] Individual conference

---

[13] USCG, *Migrant Interdiction Mission Performance Plan FY2011–2016*, July 2010d.

[14] Michel, 2014; Lee, 2014; and J. H. Korn, "District Seven FY14 Operational Planning Direction (OPD)," memorandum distributed to Coast Guard sectors and air stations, Miami, Fla.: U.S. Coast Guard Seventh District, December 27, 2013.

Table 4.3
Logic Model for Migrant Interdiction

Outcomes	Accomplishments	Activities
1. Deter undocumented maritime migration attempts	1.1. Raise public awareness of policies concerning, countermeasures against, and consequences of undocumented migration	1.1.1. Implement public-affairs campaigns in source countries in coordination with other agencies 1.1.2. Implement public-affairs campaigns domestically in coordination with other agencies 1.1.3. Publicize successful prosecutions and associated prison sentences for smugglers
	1.2. Demonstrate effective presence and the capability to deter maritime migration	1.2.1. Ensure high visibility throughout high-traffic vectors 1.2.2. Ensure periodic visibility in low-traffic vectors 1.2.3. Conduct high-profile mass-migration exercises
	1.3. Enable prosecution of smugglers	1.3.1. Identify, apprehend, and transfer smugglers 1.3.2. Prepare case packages for prosecution of smugglers
2. Prevent undocumented migrants from reaching U.S. territory via maritime routes	2.1. Build interagency and international partnerships to share information, coordinate plans, and operate cohesively	2.1.1. Engage in interagency bodies, leverage interagency centers, exercise with other agencies, and operate alongside other agencies 2.1.2. Engage bilaterally and multilaterally to enable information sharing, shipriding, training, enhancement of partner-nation capabilities, and cooperative efforts 2.1.3. Continue development of bilateral agreements to facilitate the swift repatriation of migrants after interdiction
	2.2. Achieve enhanced situational awareness of and knowledge about undocumented migration	2.2.1. Enhance abilities to predict emerging migration threats and new smuggling routes 2.2.2. Estimate migrant flow
	2.3. Detect, interdict, and repatriate migrants (or enable repatriation)	2.3.1. Detect migrants 2.3.2. Interdict migrants 2.3.3. Repatriate migrants 2.3.4. Rescue migrants from overloaded and/or unseaworthy vessels 2.3.5. Provide humanitarian aid to interdicted migrants 2.3.6. Provide access to protection screening in accordance with law, policy, and agreements

calls with the districts and migrant smuggling experts at Coast Guard Headquarters offices helped expand on the reports, memos, and metrics we assessed.

The Migrant Interdiction Mission Performance Plan FY2011–2016 notes that the Coast Guard is the lead agency enforcing U.S. immigration law at sea (although other agencies, notably Customs and Border Protection's [CBP's] Office of Air and Marine [OAM], also contribute).[15] It lists three strategic goals for migrant interdiction on maritime routes:

15 USCG, 2010d.

- deterring undocumented migrants and smugglers
- detecting and interdicting these individuals
- increasing Coast Guard participation in initiatives with a border security nexus and building relationships with agency and international partners.

Initially, we considered designating this mission as having three outcomes, one for each strategic goal but later focused on two outcomes: deterring undocumented maritime migration and preventing undocumented migrants from reaching U.S. territory via maritime routes. We recognize the critical importance of the third strategic goal for this mission but also note pursuing partnerships, relationships, and initiatives supports the other two goals. We therefore assigned parts of it to the accomplishment level in support of prevention and other, more-specific parts to the activity level in support of deterrence.

The accomplishments we developed for the logic model were heavily based on the objectives stated in the *Migrant Interdiction Mission Performance Plan*.[16] In fact, many of our objectives are slight rephrasings of those in the plan, with wording that makes them consistent with the rest of the missions' logic models. Likewise, the activities listed in the logic model were derived from performance initiatives and existing daily tasks. The elements of the logic model also align neatly with the stated priorities of the Coast Guard's *Western Hemisphere Strategy*,[17] which heavily discusses the importance of partnerships, enabling partners to build capacity, enhancing awareness, and intelligence. All these can contribute not only to border security (in both the migrant and drug contexts) but also to combating criminal networks.

Existing Metrics

Existing Coast Guard migrant interdiction metrics focus on the flow of migrants attempting to enter the United States via maritime routes and the number of undocumented migrants who were interdicted, whether by the Coast Guard or another entity. A Coast Guard performance highlights document for FY 2013 also refers to the number of smugglers detained.[18] Some metrics are calculated as percentages, dividing the number of migrants interdicted by the estimated total number. All these metrics relate to interdiction and flow.

We evaluated the validity, reliability, and feasibility of existing metrics for migrant interdiction. Overall, the majority have high validity (i.e., they effectively capture what they are intended to measure), and the remainder have medium validity. The reliability (consistency of measurement) of some metrics is limited by the fact that the total number of migrants is imperfectly known.[19] (The same problem of limited knowledge of illicit activity is inherent in other missions, including drug interdiction, LMR, and OLE.) While multiple techniques can estimate total number of migrants, the accuracy of such estimates will always be limited, and the variability among them reduces their reliability. Moreover, data on migrants who are interdicted by partner nations or other agencies may not be clearly and accurately provided to the Coast Guard, further reducing accuracy. According to experts we interviewed, there are

---

[16] USCG, 2010d.

[17] USCG, 2014d.

[18] USCG, 2014b.

[19] Some flows are better known than others. Specifically, since Cubans who reach land are eligible to remain in the United States, they have an incentive to report their arrival to the authorities. However, this is an exception. Moreover, it does not include numbers of Cubans who attempted the trip and either returned to Cuba or were lost at sea.

data discrepancies on migrants among DHS agencies and even among Coast Guard databases. These problems are exacerbated by terminology differences among DHS agencies dealing with migration, hindering their ability to fuse data.

The requirement to transfer data among multiple agencies also limits feasibility (how easily the metric can be measured) because the processes are not fully automated. Feasibility is also constrained by the resources required to generate the estimates from the limited evidence available; some approaches to estimation may require surveys, economic or demographic analyses, and even investigations of physical evidence.

Finally, in measuring asset hours for migrant interdiction, it is hard to distinguish between the simple presence of a Coast Guard asset and its activity. Because Coast Guard assets typically perform multiple missions, either simultaneously or within short time spans, this issue extends beyond the migrant interdiction mission. Our potential metrics address this issue by determining the number of asset hours by type, the time of day or night, whether the asset was "engaged" in migrant interdiction activity, and whether the asset is in a high- or low-traffic area.

We also noted the extent to which existing metrics address outcomes, accomplishments, and activities within the logic model. These metrics capture only a few items, overwhelmingly at the outcome and activity levels; existing metrics do not cover most of the outcomes, accomplishments, and activities we identified.

Potential Metrics

We developed a series of potential metrics, both to refine existing metrics and to capture all items in the logic model that are not currently being measured. Throughout this process, we kept in mind the Coast Guard's central role in migrant interdiction and that the Coast Guard needs to depend heavily on its own actions and information, with limited support and cooperation from other agencies.

Some of the potential metrics derived from existing documents in a relatively straightforward manner. For example, *Always Ready FY13 Performance Highlights* contains the following statement: "The USCG . . . detained over 190 suspected smugglers for prosecution in the United States."[20] However, neither the CGBI or other stated performance measurements included the number of smugglers detained as a metric.[21] The logic model now contains this metric and a new one relating to the ability to identify smugglers throughout the course of an interdiction. Coast Guard experts have cited the difficulty of identifying and prosecuting smugglers; measuring rates of success in this regard would be a good indicator of deterrence because the rate of recidivism for individual smugglers would speak to the success of U.S. punishment for this offense. Likewise, we did not see migrant numbers by nationality as an explicit metric, but the Coast Guard clearly captures these statistics (and publishes them online). Characterizing migrants by nationality is critical to understanding the effectiveness of particular Coast Guard efforts.

A number of the wholly new potential metrics focused on greater maritime domain awareness, which is a crucial element of success for migrant interdiction and for many other

[20] USCG, 2014b, p. 9.

[21] CGBI is an information system that collects vast amounts of data and renders that data in a searchable, analytical database to aid analytic decisionmaking.

missions. As ADM Paul Zukunft has noted,[22] there is a considerable discrepancy between what the Coast Guard can perceive and what it has the resources to be able to address; capturing that discrepancy through maritime domain awareness metrics can make the case for where additional resources would have a considerable impact.

Other potential metrics focused on measuring relatively intangible items, such as the extent or impact of public awareness (an accomplishment supporting the outcome of deterrence). The activities subordinate to this accomplishment can be measured using a number of proxies with varying degrees of validity, reliability, and feasibility: the number of public-affairs campaigns conducted in source countries, the forms of communication used by public-affairs campaigns in the United States, the estimated percentage of the audience reached, and the degree to which these campaigns influence migration decisions abroad. The last two items, on influencing migration decisions, would be the most difficult to gauge because this would likely require surveys to ensure accuracy. Surveys would be inherently time- and resource-intensive, limiting their feasibility.

Another intangible accomplishment, in support of the prevention outcome, is development of partnerships. While measuring partnerships can be subjective, we can apply such metrics as Coast Guard representation in interagency bodies, liaison personnel within other agencies or at international offices, training events with non–Coast Guard personnel, and the number of migrant exercises conducted with other agencies. Bilateral agreements and interagency memoranda are both indicators of partnerships and enablers of Coast Guard success. Collectively, these metrics contribute to an overall picture of Coast Guard partnerships, even though none wholly captures the elusive idea of "partnership."

Intelligence is also an area that is difficult to measure. One metric—its utility to users—can be assessed relatively easily, at least for tactical intelligence. When the Coast Guard Intelligence Coordination Center emails reports to the field and the broader intelligence community, it provides a link requesting feedback. The value of operational and strategic-level intelligence—for which multiple reports and analyses may contribute to a cumulative picture—may be more elusive.

Naturally, some of the potential metrics (particularly for intangible items, such as relationships) had medium or low values with respect to validity, reliability, and/or feasibility; part of the reason that some items in the logic model are not currently measured is the relative difficulty of doing so. However, a number of metrics had high values for most or all of these criteria, such as percentage of migrants interdicted by the Coast Guard who were swiftly repatriated.

Defense Readiness

The defense readiness mission supports the *National Military Strategy* and Department of Defense (DoD) operations by ensuring that Coast Guard assets are capable and equipped to deploy and conduct joint operations in support of the most critical needs of the combatant commanders in conducting major national-defense missions.[23] Coast Guard units are called on

---

[22] Presentation at the RAND Corporation, October 31, 2015.

[23] Joint Chiefs of Staff, *The National Military Strategy of the United States of America 2015*, Washington, D.C., June 2015.

to directly support the unified combatant commanders and execute essential military operations in peacetime, crisis, and war. Under U.S. Code Titles 10 and 14, the Coast Guard is defined as an armed force and is authorized to conduct the defense readiness mission.[24]

Development of the Logic Model

Table 4.4 presents the logic model for the defense readiness mission. The FY 2015 SPD, Atlantic Area OPD, and CGBI outline both specific outcomes and activities in support of the defense readiness mission.[25] Much of this mission is predicated on requests for forces emanating from DoD operational planning documents, maintaining standing homeland defense readiness requirements, historic joint documents in the form of MOUs and MOAs, and emergent national security events. Major defense readiness activities include

- maritime interception and interdiction operations
- military environmental response
- port operations, security, and defense
- theater security cooperation
- coastal sea control operations
- rotary-wing air-intercept (RWAI) operations
- combating terrorism operations
- maritime operational threat response (MOTR) support.

The defense readiness mission is unusual in that it is primarily about ensuring that the Coast Guard can support other armed forces' actions. While the Coast Guard collaborates extensively with partners in each of its statutory missions, this is the only one in which it is exclusively a supporting command rather than the supported one. In this context, all three of the outcomes we derived from Coast Guard documents are about contributing to DoD. The first outcome is to fulfill DoD requirements; at the accomplishment and activity levels, this entails providing units and conducting operations for DoD. The second is to maintain the ability to respond to calls for forces, by ensuring both readiness and interoperability with DoD. The third is to fulfill homeland defense requirements. At the accomplishment level, this includes conducting interception and interdiction operations and responding to military environmental hazards (e.g., conducting incident management, containment and cleanup, and investigations). While there has been some controversy about whether the latter is part of this mission, the *Defense Operations Program Performance Plan FY2014–2019* lists it as such and explicitly states that it is authorized by 14 U.S. Code 141.[26]

Existing Metrics

The majority of existing Coast Guard defense readiness metrics simply measure the degree to which the Coast Guard provides support in response to a request for a specific asset or capability. The metrics are typically asset centric, for example, the percentage of the time that patrol

---

[24] U.S. Code, Title 10, Armed Forces, Section 101, Definitions; U.S. Code, Title 14, Coast Guard, Ch. 1, Establishment and Duties.

[25] Michel, 2014; Lee, 2014.

[26] USCG, *Defense Operations Program Performance Plan FY2014–2019*, August 2013d; U.S. Code, Title 14, Coast Guard, Sec. 141, Cooperation with Other Agencies, States Territories and Political Subdivisions.

Table 4.4
Logic Model for Defense Readiness

Outcomes	Accomplishments	Activities
1. Fulfill DoD requirements	1.1. Provide teams and support for counterterrorism; counterdrug; and coastal, port, and harbor security operations	1.1.1. Provide visit, board, search, and seizure (VBSS) team support
		1.1.2. Provide redeployment assistance and inspection detachment (RAID) support
		1.1.3. Provide deployable force packages (e.g., K9, RWAI, Maritime Security Response Team [MSRT], Maritime Safety and Security Team [MSST])
		1.1.4 Provide airborne use of force (AUF) capability support
		1.1.5. Provide port security unit (PSU) support
	1.2. Conduct operations for combatant commanders	1.2.1. Conduct theater security cooperation (TSC)
		1.2.2. Conduct coastal air and surface operations
		1.2.3. Conduct port and harbor operations
		1.2.4. Conduct arctic and antarctic coastal sea control operations
2. Maintain ability to respond to calls for forces	2.1. Maintain readiness of forces	2.1.1 Conduct independent and joint military training
		2.1.2 Conduct independent and joint military exercises
	2.2. Maintain interoperability with DoD forces	2.2.1. Conduct joint military operations
		2.2.2. Conduct joint military communications
3. Fulfill homeland defense requirements	3.1. Conduct interception and interdiction operations	3.1.1. Execute RWAI duties
		3.1.2. Provide cutters and boats for interception and/or interdiction operations
		3.1.3. Provide law enforcement detachment team (LEDET) support
		3.1.4. Provide MOTR support
	3.2. Conduct military environmental response operations	3.2.1. Conduct spill and release notifications
		3.2.2. Conduct spill and release incident management
		3.2.3. Conduct spill and release containment and cleanup operations
		3.2.4. Conduct pollution incident investigation and documentation

boats or PSUs are available to fulfill DoD requirements. However, the majority of outcomes, accomplishments, and activities in the logic model are centered on different aspects of the mission, rather than on a specific asset. As a result, nearly all the existing metrics map to multiple elements of the logic model, limiting their validity for any one item.

One of the inherent challenges for existing defense readiness metrics is that, more than for most missions, they are driven by the demands of another department.[27] Coast Guard defense readiness reporting represents a small part of a much larger nationwide DoD-driven requirement for readiness assessments.

In characterizing existing defense readiness metrics, we recognize that these are also evolving; our documentation represents a snapshot of a changing scene. The Coast Guard continues to assess and report on its readiness for training and operations; for the larger cutters, these activities are fully integrated into the USN afloat readiness training regime and are currently referred to collectively as "tailored ship's training availability." A recent reorganization and streamlining of these functions within the Coast Guard's Force Readiness Command enterprise was not complete as of this writing but was nearing final operating capability. Shore forces and smaller cutters conduct ready-for-operations activities and assess special emergency operations and procedures readiness. Reference to the Status of Readiness and Training System (SORTS) metrics and reporting has been phased out and replaced with the Defense Readiness Reporting System, rendering some of the existing metrics obsolete. Defense Readiness Reporting System metrics are being integrated into the emerging Coast Guard Resource and Capabilities Evaluation System (CG-RACE), which will present additional defense readiness metrics for consideration.

Potential Metrics

We intend for some classes of our potential metrics to capture the somewhat elusive ideas of interoperability and readiness. Proxies relating to participation in joint exercises or training events are highly feasible but have medium validity. On the other hand, ratings of interoperability and readiness can be garnered by polling relevant DoD and Coast Guard personnel, although this entails limited reliability (due to subjectivity) and low feasibility (due to the difficulty of collecting the data). Exit interviews from personnel leaving joint commands can contribute to data collection, but that information still needs to be captured and integrated.

Other potential metrics relate to activities supporting military environmental response operations, such as notification, cleanup, and investigations. For some of these activities, ascertaining the Coast Guard's performance is inherently difficult; many factors affecting the impact and cleanup of a spill are beyond the Coast Guard's control. Nonetheless, we have listed some potential metrics, acknowledging and documenting their limitations.

In a number of cases, we have added "percentages of requests met" for additional types of support that DoD may request and that are not enumerated in existing metrics. We have also proposed measuring some outcomes and accomplishments using weighted averages of metrics for subordinate elements of the logic model.

Summary of Findings on Security Missions

As with the safety missions, existing metrics do not capture many elements of the logic models for the four security missions (PWCS, drug interdiction, migrant interdiction, and defense

[27] This occurs to some extent in other missions—for example, the cocaine-centric nature of drug interdiction metrics reflects other agencies' focus—but is more pronounced for this mission.

readiness); see Table 4.5. As before, this simply reflects the fact that the metrics were developed prior to and independently of the logic models.

Of the existing metrics, 82 percent were highly feasible (easy to measure), and 74 percent were highly reliable. For three of the missions, all or nearly all of the existing metrics also had high validity. However, as noted previously, there was a fundamental misalignment between the defense readiness mission's asset-centric existing metrics and the mission-centric logic model, so metrics did not capture the elements of the logic model well. This is not meant as a critique of those who developed the metrics but as an indication that alternative metrics may be needed to measure mission performance. (The existing metrics may have value in other contexts, such as resource allocation, although we did not analyze this.) The potential metrics for the defense readiness mission can help to fill these gaps, just as potential metrics for other missions can help expand coverage of those missions' logic models and refine select metrics as needed.

Table 4.5
Numbers of Security Mission Logic Model Elements and Percentages Measured by at Least One Metric

Type of Element	Number	Percentage
Outcomes	9	77
Accomplishments	21	38
Activities	72	31
Total	102	36

CHAPTER FIVE

Stewardship Missions

This chapter discusses the logic models, existing metrics, and potential metrics for the four missions under the heading of stewardship: ice operations, ATON and WWM, MEP, LMR, and OLE. As in the previous two chapters, we describe each mission in one or two sentences, then present and discuss its logic model. Next, we briefly characterize existing metrics, including their evaluations in terms of validity, reliability, and feasibility. Finally, we highlight key aspects of the potential metrics we identified. As before, comprehensive tables of the existing and potential metrics can be found in the appendixes.

Ice Operations

The ice operations mission involves enabling movement and access despite icy conditions. Most of this mission involves domestic icebreaking to enable maritime commerce, particularly on the Great Lakes (District 9) and along the northern portions of the eastern seaboard (District 1 and, to a lesser degree, District 5). Other components of the mission include polar icebreaking, rescuing beset vessels, and clearing ice jams.

Development of the Logic Model

The logic model for the domestic icebreaking mission (Table 5.1) is designed to connect the Coast Guard's icebreaking-related prevention, monitoring, regulatory, and response activities with the desired outcomes, which are to facilitate commerce while saving life and property despite icy conditions. The content of the logic model came, directly or indirectly, from discussions of activities and objectives in the following documents: the FY 2015 SPD, Atlantic Area Minimum Mission Standards, the Assistant Commandant for Prevention Policy's *Marine Transportation Systems Management Program Performance Plan*, and current extracts from OPARs.[1] Discussions with representatives with Coast Guard area and district personnel for whom ice is geographically relevant allowed further refinement of the model.

Most of the assets devoted to this mission are used for activity 1.1.3, breaking ice to enable movement on "Tier I" (high-priority) waterways in the northeastern United States and Great Lakes. This activity allows use of these waterways to handle large-scale commercial traffic (such as bulk materials) and to provide isolated communities with supplies (such as heating oil). However, other, less-visible responsibilities also fall under the rubric of ice operations. The

[1] Michel, 2014; Lee, 2014; USCG, "Atlantic Area Minimum Mission Standards," undated a; Joseph Servidio, *U.S. Coast Guard Marine Transportation Systems Management Program Performance Plan, Fiscal Years 2014–2019*, Washington, D.C.: U.S. Coast Guard, August 2013a, Not available for public release.

Table 5.1
Logic Model for Ice Operations

Outcomes	Accomplishments	Activities
1. Facilitate safe and efficient activity in icy waters while reducing the risk and severity of ice-related incidents	1.1. Enable maritime commerce to proceed safely without significant delay despite icy conditions	1.1.1. Monitor and report ice conditions on coastal and inland waterways 1.1.2. Ensure that vessels operating independently in ice have sufficient horsepower and hull reinforcement[a] 1.1.3. Break ice to allow certain commercial vessels to utilize iced waterways as necessary 1.1.4. Coordinate with other public and private icebreaking entities to ensure adherence and appropriate modifications to standards
	1.2. Enable access to icebound polar regions	1.2.1. Deploy polar-class icebreakers to polar regions
	1.3. Minimize human casualties, property damage, and environmental damage from vessels beset by ice	1.3.1 Rescue vessels beset by ice
	1.4. Minimize human casualties and property damage from flooding due to ice jams	1.4.1. Break ice jams that pose a flooding hazard

[a] This has some overlap with the marine safety mission.

Coast Guard's monitoring, reporting, and oversight responsibilities are captured in other activities subordinate to accomplishment 1.1. The Coast Guard also conducts polar icebreaking, as indicated by accomplishment 1.2.[2] The importance of this accomplishment is underscored by the Coast Guard's *Arctic Strategy* (2013c), which discusses access issues in the region. As the last two accomplishments indicate, the Coast Guard also rescues vessels beset by ice and breaks ice jams that pose a flooding hazard.

Existing Metrics

Existing Coast Guard metrics for ice operations focus on enabling maritime commerce in domestic waterways, relating to activity 1.1.3. These metrics include counts of the number of icebreaking requests met and the number of days of cutter availability. In addition, the percentage of time that high-priority waterways were open was also tracked. The Coast Guard recently changed its metrics for measuring how well it is facilitating commerce. Instead of measuring the number of days the waterway was impeded, it now measures the number of days the waterway was open.

We evaluated the existing metrics as high in terms of validity, reliability, and feasibility. The data on requests for assistance, waterway availability, and cutter availability are easy to collect and very directly reflect the Coast Guard's level of activity for enabling commerce and assisting vessels.

[2] USCG, 2013c, underscores importance of this accomplishment by discussing access issues in the region.

Potential Metrics

Several new metrics could help the Coast Guard better assess its progress in meeting all the outcomes specified in the logic model. The first is a measure that speaks to the core of the mission: the economic impact of the mission itself, normalized by the ice severity. To determine this, the Coast Guard could either collect data about the vessels assisted or use statistical data from the Army Corps of Engineers or the Department of Commerce (e.g., Bureau of Economic Analysis, Bureau of Labor Statistics, and the National Weather Service). These agencies are likely to offer their assistance as a means of testing a potential new approach to measuring the economic impact of government activities. Despite these data sources, evaluating the economic impact of icebreaking may face some difficulties, which is why the feasibility and reliability of this metric are medium. Calculating the degree to which economic effects propagate (i.e., how much a delivery affected the downstream economy) requires at least limited economic modeling. Measuring the economic effects of heating-oil deliveries when they are needed for human safety is also complex.

The economic impact of the assistance rendered should be considered in the context of the severity of the ice. This is because thicker and/or more severe ice conditions make rendering assistance much more time consuming and expensive.[3] Ice conditions may vary greatly from one year to the next, as well as among subseasons and different locations; normalizing by ice severity can help to make these differences apparent. However, it may also introduce some uncertainty and subjectivity, reducing reliability and feasibility for metrics in which this normalization is used.

Another aspect of ice operations is responding to maritime incidents that occur as a result of icy conditions. In the ATON and WWM section, we propose including SAR incidents in places in which collisions, allisions, and groundings (CAGs) are measured.[4] For domestic ice operations, we also suggest tracking other maritime mishaps in which the ice or other winter season conditions are a factor (CAGs, SAR, and mishaps [CAGSMs]). (Determining whether particular actions would have prevented or mitigated particular CAGSM incidents still involves some degree of judgment.) As we will also suggest for ATON and WWM mission metrics, denominators, such as vessel-miles, ton-miles, dollar-miles, or passenger-miles, can be used to reflect underlying waterway usage. By calculating the CAGSM rate per waterway usage in terms of vessel-miles traveled, the Coast Guard can evaluate success with respect to the first outcome (enabling vessel activity while minimizing incidents). Again, this should be normalized by the severity of the ice in the environment.

The rest of the proposed measures for the ice operations mission have a similar pattern: They attempt to measure the number of undesirable situations observed or reported and to put them into the context of the severity of the ice season and/or the economic activity in the area. We complement existing metrics by covering aspects of the logic model that they do not, such as rescue of beset vessels, breakup of ice jams, and polar icebreaking.

---

[3] The severity of icy conditions is not solely a function of the thickness of the ice. Ridging and other complex ice phenomena can hinder icebreaking or movement through broken ice. Conditions may also be correlated with greater need. In some cases, such as the delivery of heating oil, a more-severe winter can also make the assistance rendered much more urgent, since the oil is consumed more quickly in colder weather.

[4] An *allision* is when a moving object hits a stationary one, while a *collision* is between two moving objects.

Aids to Navigation and Waterways Management

According to the Assistant Commandant for Prevention Policy's *Marine Transportation Systems Management Program Performance Plan*, the purpose of this mission is to ensure "a safe, secure, efficient and environmentally sound waterways system."[5] This includes, but is not limited to, the statutory mission of providing and maintaining ATON.

Development of the Logic Model

The logic model for the ATON and WWM mission (Table 5.2) reflects the dual nature of the mission, promote safety and facilitate commerce. This mission includes such programs as WWM, ATON, marine event permitting, and regulation of bridges over navigable waterways. The content of the logic model was drawn from precisely the same documents that were listed above as having been used for the ice operations logic model. Discussions with representatives in each Coast Guard area and district helped further refine the model.

Existing Metrics

Most existing metrics focus on monitoring, placing, and maintaining ATON. This makes sense because significant resources have been invested in the personnel, equipment, and vessels that perform most ATON system maintenance. One metric in particular—the number of CAGs—has previously been construed as reflecting the deficiencies of the ATON system.

Table 5.2
Logic Model for Aids to Navigation and Waterways Management

Outcomes	Accomplishments	Activities
1. Facilitate the safe and efficient use of the Marine Transportation System (MTS)	1.1. Reduce impediments to navigation	1.1.1. Regulate the placement and operation of bridges and offshore infrastructure over and in navigable waterways to minimize barriers to navigation
		1.1.2. Coordinate with the U.S. Army Corps of Engineers and other entities to provide input on the regulation and removal of temporary and permanent potential obstructions to navigation
		1.1.3. Maintain up-to-date Waterway Analysis and Management System (WAMS) for planning
	1.2. Achieve high levels of mariner awareness of navigational conditions and environment	1.2.1. Regulate and plan the optimal placement and maintenance of public and private navigation infrastructure
		1.2.2. Place, maintain, and (as necessary) remove navigational infrastructure
		1.2.3. Provide the maritime public with information on discrepancies in navigation infrastructure and other unpublished changes to maritime navigation safety
	1.3. Reduce frequency of maritime accidents resulting from special circumstances	1.3.1. Establish, maintain, and monitor limited-access areas
		1.3.2. Regulate maritime events
		1.3.3. Direct maritime traffic

[5] Servidio, 2013a.

However, SMEs indicated that, given modern navigational technologies (such as the Global Positioning System) and redundancies among ATON, ATON discrepancies likely contribute to only a fraction of the total number of CAGs. Holding the ATON system "accountable" for all CAGs leaves out other key contributing factors, such as human error and mechanical malfunctions within vessels.

A number of very detailed metrics are also associated with aspects of the ATON system. These include the availability of aids by priority, timeliness of maintenance schedules, and total numbers of ATON components by type (submarine cables, sound signals, fog detectors, etc.). Notable among these was a measure to improve the accuracy of ATON reports. The target of this measure was to reduce the "reasons not reported" for ATON discrepancy reports to less than 5 percent. There are also metrics related to the processing time for bridge permits and regulations, marine event permits, and whether WAMS is up to date.

We evaluated the existing metrics in terms of validity (how well they capture the element they measure), reliability (how consistently measurements can be made), and feasibility (how easily measurements can be made). Reliability and feasibility are high for nearly all the existing metrics. For the CAG metric, validity is low because many CAGs have no relationship to the existence or availability of ATON, but reliability is medium because not all CAGs, especially minor ones, are reported to the Coast Guard. Other metrics rated as medium included those that measure administrative processing times because adjustments in processing times would have a marginal affect on the navigability of waterways that are affected by the items that are being regulated.

Potential Metrics

We propose several new metrics to help the Coast Guard better assess its progress in meeting all the outcomes specified in the logic model. Instead of using the original CAG metric, we include collisions, allisions, groundings, *and* SAR cases (CAGSs). (In the previous section of the paper, we had included collisions, allisions, groundings, SAR cases, and mishaps in ice under the rubric of CAGSMs; however, mishaps in ice are not relevant to the current mission.) We propose to use only the CAGS cases *in which ATON is a factor* as the numerator of a new metric. The denominator should measure waterway usage in terms of vessel-miles, ton-miles, dollar-miles, or passenger-miles, depending on the context and on what data sets can be most feasibly collected. For example, vessel-miles may be appropriate if the Coast Guard wants to evaluate usage without reference to vessel size, while ton-miles would capture the fact that vessel sizes vary. Using dollar-miles would underscore cost-effectiveness issues, while using passenger-miles would focus on the number of people being moved via ferry, cruise ship, or recreational vessel. The same choice of denominator recurs for several other proposed metrics. For the most part, the requisite data will be collected by local ports, the Army Corps of Engineers, or the Bureau of Economic Analysis.

Among the possible new metrics for the outcome of maintaining unimpeded, navigable waterways are the overall satisfaction of waterway users and number of user complaints. If implemented, the data for these metrics would be collected by periodically surveying the public or specific waterway user communities. Validity and reliability were generally higher for these types of metrics. However, in terms of feasibility, implementing a broad survey of waterway users can be difficult. At least one DoD agency has simplified the process of widespread customer satisfaction surveys by appending a link to a feedback survey at the end of every email the agency sends.[6]

The other proposed metrics generally identify aspects of the activity that can be measured and then attempt to place the potential measures in an appropriate context. For measures involving waterway usage, as indicated above, the Coast Guard can use vessel-miles, ton-miles, dollar-miles, or passenger-miles, as appropriate. For other metrics, such as permits for marine events or bridges, we suggest normalizing according to the number and complexity of applications. For limited-access areas, we suggest normalizing according to how well the need for public notice was balanced against the need for immediate action. Many of these potential metrics have medium or high validity, reliability, and feasibility, indicating that we think that the metrics could reasonably be collected and would provide important and useful insight into mission accomplishment.

Marine Environmental Protection

The MEP program is highly complementary with the marine safety program, discussed in Chapter Three. Like marine safety, MEP involves some element of inspections to reduce risk. However, its main emphasis is on ensuring preparedness for environmentally hazardous spills and on responding to such spills when they occur.

Development of the Logic Model

Using the same sources as for marine safety, we designated three broad outcomes for MEP (Table 5.3). The first, decreasing the probability and prospective impact of spills, entails three accomplishments: increasing responder capabilities, improving response plans, and reducing risks. The supporting activities include outreach, exercises, assisting in the development of plans, and inspections.

The second outcome involves responding to spills. At the accomplishment level, the Coast Guard has the responsibility, as the federal on-scene coordinator, to ensure that oil spills and hazardous substance releases within its jurisdiction are contained and remediated. The other accomplishment relates to legal investigations (including citations at the activity level).

Finally, the third outcome focuses on a lesser-known aspect of this mission: monitoring incident hazards and communicating them to the public. This requires a mix of technically demanding modeling and effective, clear outreach to people who may be affected.

Existing Metrics

Several closely related MEP metrics address exercise completion rates, which help to address responder proficiency and coordination. Some focus on numbers of spills and recovery rates of spilled material. A key issue with these metrics is that they are shaped by many factors beyond the Coast Guard's control, some of them incident specific. Others focus on the percentages of spill reports that resulted in a Coast Guard response or the percentage of cases in which an after-action report (AAR) was submitted. We also discovered an unusual one, without any counterpart that we were able to identify for other missions: the MEP "efficiency ratio," which incorporated both numbers of spills and the cost of the MEP program.

---

[6] RAND observed this invitation for feedback during interactions with the Army National Guard Bureau.

Table 5.3
Logic Model for Marine Environmental Protection

Outcomes	Accomplishments	Activities
1. Decrease probability and prospective impact of spills and releases	1.1. Increase proficiency of responders and coordination with them	1.1.1. Conduct unit, responder, and industry training and awareness and maritime community outreach efforts
		1.1.2. Conduct interagency, state, local, and private-sector spill and disaster preparedness exercises, including government-initiated unannounced exercises (GIUEs)
	1.2. Improve response plans	1.2.1. Assist with development of industry, state, and local response plans
	1.3. Reduce risk from vessels and facilities	1.3.1. Conduct safety and International Convention for the Prevention of Pollution by Ships (MARPOL) examinations of vessels and facilities
2. Mitigate effects of spills and releases through response operations	2.1. As federal on-scene coordinator, ensure that spills and releases are contained and remediated	2.1.1. Monitor and document cleanup and remediation operations, supervise contractors, and coordinate among agencies
		2.1.2. Deploy pollution first responders and respond to incident
	2.2. Conduct legal investigation of spill and release incidents	2.2.1. Investigate incidents by documenting evidence, pollution sources, causal factors, and responsible parties, while also contributing to others' investigations
		2.2.2. Prepare for legal proceedings by issuing appropriate citations and prepare case packages for prosecution
3. Monitor and communicate hazards to the public	3.1. Monitor and model flow of hazardous substances	3.1.1. Detect and track movements of hazardous substances
		3.1.2. Model the anticipated movements of hazardous substances
	3.2. Inform the public to reduce exposure to hazardous substances	3.2.1. Designate evacuation and exclusion areas based on current and anticipated hazards
		3.2.2. Disseminate information on evacuation and exclusion areas

All the existing MEP metrics have high feasibility and high or medium validity and reliability (with the exception of the MEP efficiency ratio, which we could not map to an element of the logic model). The main issue is not the quality of existing metrics, but their limited coverage: The nine metrics measure only four of the 22 elements of the logic model, those involving exercises and cleanup—its most visible and most easily measured aspects. Less-prominent and harder-to-measure aspects of the mission, such as aiding in the development of plans and monitoring the flow of hazardous substances, do not have existing, widely disseminated metrics.

Potential Metrics

The potential MEP metrics we developed fall into several broad categories. Some are meant to examine very specific aspects of preparedness, for example, the percentage of people or entities exceeding a certain level of training or exercise performance, complementing existing metrics that measure numbers of exercises. Unfortunately, these tend to have low feasibility, since they require surveys, quizzes, or live-action tests. It may also be challenging to standardize the level of difficulty of the exercises or tests; particularly challenging exercises may be more valuable than easier ones in some cases but may seem to indicate lower levels of performance. Likewise, some entail the use of models to measure risk reduction or economic costs, which also requires considerable effort (at least at the outset), limiting feasibility.

Other metrics focus on aspects of investigations, such as completion, citation, and successful prosecution rates and on the percentage of cases in which causal factors or pollution sources are determined. Some of these are at least partly outside the Coast Guard's control, limiting validity.

Several of the metrics aim to measure how well the Coast Guard does at modeling and communicating about hazards. This set involves estimation, surveys, and judgment to be able to characterize how well the Coast Guard performed, limiting the reliability and feasibility of these metrics. Presumably, periodically revisiting how best to make these measurements could reduce some of the subjectivity over time; however, personnel turnover and changing conditions may counter this by increasing variability. (The same issues apply for other missions.)

Overall, many elements of this mission are difficult to measure. This is reflected in the fact that few of the potential metrics for this mission had high values across all three criteria, and many had low values in one or more.

Living Marine Resources

The Coast Guard's LMR mission is intended to enforce domestic fisheries laws at sea.

Development of the Logic Model

We drew from several sources in developing the logic model for the LMR mission (Table 5.4), the most important of which was *Ocean Guardian* in both its 2004 and 2014 incarnations.[7] Additionally, we relied on the *Living Marine Resource Enforcement Performance Plan* of July 2010,[8] the FY 2015 SPD,[9] and conversations and interviews with Coast Guard personnel from the districts responsible for most LMR mission activities. These conversations added critical insights and detail to our understanding of the mission and how best to measure its success. These personnel also repeatedly stressed the importance of *Ocean Guardian* and the extent to which it provides guidance, philosophy, and metrics for the LMR mission.

The *Living Marine Resource Enforcement Performance Plan* notes that the strategic goals of the LMR mission are protecting U.S. LMR, maintaining a level playing field for U.S. fishing, and supporting efforts to reduce threats to endangered species.[10] *Ocean Guardian's* strategic goals refer to protection of natural resources and maritime security; the guidance provided by the SPDs indicated that fishing regulation compliance rates had high priority. We integrated

---

[7] USCG, *Fisheries Enforcement Strategic Plan: Ocean Guardian*, Washington, D.C., September 2004, and USCG, *Fisheries Enforcement Strategic Plan: Ocean Guardian*, Washington, D.C., 2014a. Although the 2014 version supersedes the 2004 version, we found that the earlier version provided useful information that we wanted to draw on.

[8] J. E. Ryan, *Living Marine Resource Enforcement Performance Plan, FY 2011–2016*, U.S. Coast Guard, July 1, 2010, Not available to the general public.

[9] Michel, 2014.

[10] Ryan, 2010.

Table 5.4
Logic Model for Living Marine Resources

Outcomes	Accomplishments	Activities
1. Prevent illegal fishing activities and other threats to marine species	1.1. Communicate and work with both other agencies and the general public	1.1.1. Continue and expand partnerships with other agencies
		1.1.2. Share information with the public
	1.2. Enforce regulations	1.2.1. Maintain an effective presence to achieve awareness and deter prospective violations
		1.2.2. Monitor for and detect overt LMR violations (e.g., based on location, timing, activity, gear)
		1.2.3. Detect and deter less-visible LMR violations by conducting random and/or targeted boardings
		1.2.4. Respond to LMR violations—document, disrupt, intercept, interdict, board and/ or apprehend, process as appropriate
		1.2.5. Cite violators and prepare cases for prosecution, as appropriate
	1.3. Contribute to conservation efforts	1.3.1. Monitor, collect, analyze, and disseminate information about specific marine species
	1.4. Ensure Coast Guard compliance with laws and regulations for LMR	1.4.1. Engage with relevant Coast Guard units and programs to ensure knowledge of and adherence to laws and regulations

these strategic goals into one main outcome for the LMR logic model: to prevent illegal fishing activities and other threats to marine species.[11]

The Coast Guard has several avenues for executing this mission. The main avenue is enforcement of regulations, mentioned in all planning documents and conversations with personnel. Enforcement heavily informs all three of *Ocean Guardian's* program goals: (1) prevent illegal encroachment of the exclusive economic zone (EEZ) by foreign fishing vessels (FFVs),[12] (2) effectively enforce federal regulations, and (3) ensure compliance with international regulations.[13] Enforcement entails boardings, interdiction, citation of violators, maintaining presence, and monitoring to prevent illegal fishing, which our interviews indicated is more of a problem in some districts than in others.

Ocean Guardian and the FY 2015 SPD also refer to communication and outreach as means of pursuing this outcome.[14] The Coast Guard works in concert with other government bodies on the LMR mission, including the Department of State and the National Oceanic and Atmospheric Administration (NOAA), and must also ensure that the public—at least, the sector of the public engaged in fishing or related activities—is aware of current law and regulations. Partnerships with other agencies and information-sharing campaigns further these goals.

---

[11] USCG, 2004; USCG, 2014a.

[12] The EEZ extends 200 nmi from shore. Where the 200-nmi zones of two nations overlap, a median line is typically used to demarcate EEZs.

[13] USCG, 2004; USCG, 2014a.

[14] USCG, 2004; USCG, 2014a; Michel, 2014.

The conservation portion of the LMR mission comprises Coast Guard efforts to monitor, collect, analyze, and disseminate information on specific marine species. An example is the effort to inform large ships about the endangered northern right whale environment in the Atlantic; the Coast Guard works with NOAA, DoD, the Marine Mammal Commission, and other agencies to ensure that large vessels traveling through areas inhabited by northern right whales are aware of conservation efforts in the area. Finally, the Coast Guard must ensure that its own personnel comply with existing laws and regulations on fisheries and conservation.

Existing Metrics

Current Coast Guard LMR metrics are mostly articulated in terms of percentages, of boarding rates, and of surveillance of high- and low-threat areas, as would be expected for measurement of regulation enforcement. Three metrics measure absolute numbers of requests for assistance and education campaigns about conservation. However, the existing metrics do not measure other, crucial parts of the LMR mission, such as communication with the public and with other government agencies tasked with parts of this mission (such as NOAA or the Department of State), and the Coast Guard's own compliance with LMR laws and regulations.

Our evaluation of the validity, reliability, and feasibility of existing metrics revealed that all rated as high or medium across all three categories. The majority of metrics rated as high in feasibility. About one-half rated as medium in reliability, mainly because, for these metrics, the denominator may not be accurately known. For example, the exact number of fishing vessels involved may not be known to calculate a boarding rate. While the Coast Guard may estimate the denominator based on the number of vessels sighted or detected, there will always be some degree of inaccuracy, particularly for districts encompassing wide geographic areas.

Two existing metrics that measure absolute numbers of a task or activity—specifically, pulse operations and outreach campaigns for protected species—were assessed as having medium validity. It is unclear that the number of pulse operations for protected species is comparable across districts; protected species, fishery activity, and geographic extent differ greatly among districts, which would seem to preclude meaningful comparisons.[15]

Two other existing metrics—the percentage of requests for LMR conservation assistance met and the percentage of fisheries with satisfactory enforcement levels—score high across all three of our criteria. In addition to being reasonable methods of assessing the Coast Guard's level of involvement in LMR conservation and enforcement, these metrics have denominators that are readily available.

Potential Metrics

Our potential metrics address areas of the logic model for which outcomes, accomplishments, or activities are not currently measured and refine some existing metrics. As noted, while the Coast Guard maintains responsibility for a large part of the LMR mission, it must also work in concert with other government bodies and agencies and with international partners. Measurements of Coast Guard performance must therefore take into consideration the fact that success or failure with this mission depends, to a certain extent, on factors outside the Coast Guard's direct control.

---

[15] Although we did not evaluate numerical requirements for specific metrics, the OPAR requirement for one protected species pulse operation and one outreach campaign per district per year is very consistent, given the disparities among the districts.

We have provided many potential metrics for communication and outreach, to complement the existing requirement that each district conduct at least one outreach campaign annually. Some of these are drawn from *Ocean Guardian's* discussion of Coast Guard partnerships; the activities mentioned (participation in industry trade shows to present the LMR mission, walking the docks, etc.) suggested different methods of keeping track of their effectiveness.[16] While "public awareness" may be an ambiguous goal, it is nevertheless possible to characterize public information campaigns and forms of communication, to estimate the percentage of target audiences such campaigns reach, and to estimate the percentage of the target audience actually learning from the campaigns. Other outreach measurements might include the number of high-level meetings with interagency and international partners, numbers of liaison personnel addressing LMR issues at other agencies, and numbers of Coast Guard shipriders aboard vessels belonging to other agencies and taking part in LMR activities. Many of these new metrics have medium or high validity, reliability, and feasibility. Some of them, however, do rate low on feasibility; this group comprises the metrics that would require surveys or quizzes.

The effectiveness of partnerships can be difficult to capture. In many cases, the more natural and organic a partnership is, the fewer measurable footprints it will leave behind. An interagency group of analysts speaking to one another several times a day will likely be more productive than a formal liaison group whose members rarely speak apart from designated meetings, although a close partnership will be less readily measured unless there is an intrusive count of emails or telephone calls. However, some of the potential metrics, such as the number of multiagency LMR exercises, measure activities that do further relationships.

We have made several suggestions for enforcement metrics, most of which are intended to capture more elements within the logic model. The existing enforcement metrics score highly in the three areas of validity, reliability, and feasibility, but the new ones capture additional data on detection of LMR violations, how many of these detections lead to interceptions, and how many interdictions result in penalties. If the Coast Guard adopts some or all of these metrics, a more complete picture of LMR mission effectiveness over the entire enforcement life cycle should emerge.

The enforcement accomplishment does contain some elements that are difficult to measure. As one interviewee observed, imposition of penalties does not necessarily indicate success of the LMR mission: If all fishing vessels follow all rules and if no penalties are imposed, the scores for these metrics would be zero, but the mission would be successful.

Other potential enforcement metrics measure the probability of detection of overt LMR violations, estimated percentages of vessels committing overt violations, and other estimates that would require modeling. These metrics are therefore rated as low on feasibility because of the expense and effort involved in such modeling. However, if adopted, they could provide valuable data on violations, particularly in districts where the area of operations is too large to realistically monitor.

Finally, we made two suggestions for measuring Coast Guard compliance with LMR regulations. Both would require surveys, limiting feasibility, but these metrics are intended to measure elements that are not currently being addressed.

---

[16] USCG, 2004; USCG, 2014a.

Other Law Enforcement

The Coast Guard's OLE mission prevents illegal FFV encroachment in the U.S. EEZ and enforces international agreements intended to suppress illegal, unreported, and unregulated fishing activity on the high seas.

Development of the Logic Model

We drew from a number of sources in developing the logic model for the OLE mission (Table 5.5). The most important of these were *Ocean Guardian,* in both its 2004 and 2014 incarnations, and the 2010 *Living Marine Resource Enforcement Performance Plan.*[17] Additional sources included the 2014 *Maritime Response Program Performance Plan* and FY 2015 SPD.[18] We also spoke to Coast Guard personnel involved with the OLE mission in several districts, who provided additional detail. All these personnel emphasized the importance of *Ocean Guardian* in providing guidance and metrics for the OLE mission.[19]

The *Living Marine Resource Enforcement Performance Plan* characterizes OLE goals as ensuring the integrity of the EEZ, deterring illegal fishing in areas of U.S. jurisdiction, curtailing illegal fishing on the high seas, and advancing U.S. interests through international agree-

Table 5.5
Logic Model for Other Law Enforcement

Outcomes	Accomplishments	Activities
1. Enforce U.S. EEZ	1.1. Deter illegal foreign fishing in the U.S. EEZ	1.1.1. Maintain an effective presence to deter prospective EEZ violations
		1.1.2. Publicize the risks and costs of committing EEZ violations
	1.2. Counter illegal foreign fishing in the U.S. EEZ	1.2.1. Monitor for and detect foreign ships illegally fishing in the U.S. EEZ
		1.2.2. Intercept and interdict FFVs
		1.2.3. Board, apprehend, collect evidence on, and cite FFVs
		1.2.4. Prepare case packages for prosecution, as appropriate
2. Enforce adherence to international fishing regulations	2.1. Prevent violations of international fishing regulations	2.1.1. Continue and expand effective partnerships to improve partner capabilities and information sharing
		2.1.2. Maintain an effective presence to achieve awareness and to deter prospective violations of international regulations by demonstrating the ability to enforce them
		2.1.3. Document and respond to violations of international fishing regulations—disrupt, intercept, interdict, board, apprehend, cite, and/or prepare case packages for prosecution, as appropriate

[17] USCG, 2004; USCG, 2014a; Ryan, 2010.

[18] Peter J. Brown, *United States Coast Guard Maritime Response Program Performance Plan, Fiscal Years 2014–2019,* Washington, D.C.: U.S. Coast Guard, August 2013, Not available to the general public; Michel, 2014.

[19] USCG, 2004; USCG, 2014a.

ments and enforcement of international fishing regulations.[20] *Ocean Guardian* identifies its strategic goals as protection of natural resources and maritime security.[21]

In light of these documents, we identified two outcomes for the Coast Guard's OLE mission: the enforcement of the U.S. EEZ (with two subordinate accomplishments) and enforcement of adherence to international fishing regulations (with one accomplishment). For EEZ enforcement, from *Ocean Guardian's* program goal of preventing illegal encroachment of the EEZ by FFVs, we identified such activities as maintaining effective presence to achieve awareness and deter violations, publicizing the risks and costs of EEZ violations, monitoring for and detecting incursions, and documenting and responding to EEZ violations.[22] The Coast Guard's enforcement of international fishing regulations is also achieved by maintaining an effective presence to deter violations, documenting and responding to violations of international fishing regulations, and maintaining and expanding partnerships with other agencies and international partners. *Ocean Guardian* especially emphasizes that such partnerships are crucial to the success of the OLE mission, naming the Departments of State and Commerce, individual states, and resource user groups (such as recreational groups, nongovernmental organizations, and commercial groups) as existing or prospective partners.[23] While partnerships contribute to both OLE outcomes, they are more critical to the international fisheries outcome and have been enumerated accordingly.

Existing Metrics

Coast Guard OLE metrics currently measure activities intended to prevent illegal foreign fishing in the EEZ. The FY 2014 SPD discusses performance outcome measures, such as the detection, interception, and interdiction rates for FFV incursions in the EEZ.[24] We were unable to locate any existing OLE metrics that capture data on effective presence, responses to violations of international fishing regulations, or partnerships.

We rated some of these metrics (interception and interdiction rates for detected FFVs) as high for feasibility and reliability but medium for validity. The interception rate for detected FFVs currently includes interceptions conducted by parties other than Coast Guard personnel, but the information passed on by other organizations can be incomplete. Other metrics were rated low in reliability because of the difficulty of accurately assessing the number of actual EEZ violations taking place (the recurring "unknown denominator" problem seen in other missions). As noted in our evaluation of the LMR metrics, the number of violations is very hard to estimate with any accuracy in some cases, particularly for geographically dispersed districts, such as District 14 (Hawaii, Guam, and other Pacific islands) and District 17 (Alaska).

Potential Metrics

As with the LMR mission, our potential metrics for the OLE mission refine the existing metrics and address elements of the logic model that are not yet measured. Not surprisingly, some

---

[20] Ryan, 2010.

[21] USCG, 2004; USCG, 2014a.

[22] USCG, 2004; USCG, 2014a.

[23] USCG, 2004; USCG, 2014a.

[24] Peter V. Neffenger, "Fiscal Year 2014 Strategic Planning Direction (SPD)," memorandum, Washington, D.C.: U.S. Coast Guard, September 13, 2013b, Not available to the general public.

of these elements are not yet measured because it is very hard to do so; these include both outcomes. While we have developed potential metrics estimating the economic losses and numbers of species with significantly diminished numbers due to violations of the U.S. EEZ or international regulations, we understand that these assessments require models and assumptions, rely heavily on information from other agencies, and are heavily affected by natural and human forces beyond the Coast Guard's control.

We have revised some existing metrics for countering EEZ violations, using a more readily assessed denominator. Four new metrics capture the differences between having visible assets present in an area and monitoring an area through other means. Knowing what difference these methods make (if any) could inform future monitoring efforts and would make a valuable addition to arguments for increased maritime domain awareness resources. Six metrics relate to enforcement, determining the percentages of one step of enforcement that are successful enough to enable the next. We also include metrics measuring the percentage of detections that lead to penalties. While the LMR mission does not demand interdiction or penalties for every detected vessel, the OLE mission does: *Ocean Guardian* states that the standard for prevention of illegal encroachment of the EEZ involves detection of and response to all violations.[25] Some of the metrics for EEZ violations (e.g., the percentage of successful prosecutions) depend on a number of factors and personnel outside the control of the Coast Guard. However, it would nevertheless be useful to know the values of such metrics; for example, a low rate of successful prosecution could be further analyzed to determine whether the handling of evidence or case packages was a contributing factor.

We also provide several potential metrics for enforcing international fishing regulations, many of which are rated as high in reliability and feasibility but medium in validity. A number of these concern Coast Guard partnerships. The Coast Guard must work with a wide variety of U.S. and international partners to successfully execute the OLE mission, particularly the international fisheries outcome. These metrics are generally expressed as percentages: the actual number of items (meetings, agreements, liaised personnel) divided by the desired number of items. Partnership metrics constructed in this way allow the Coast Guard to track progress and to easily adjust the metric for changed circumstances. Each of these metrics provides a data point that, on its own, may not give a complete picture of partnership productivity. Collecting the data, however, could give valuable insight on areas where partnerships could be strengthened.

We have also provided metrics for maintaining effective presence, centering on the percentage of the time that the Coast Guard has a visible presence in or can monitor a region. It is quite likely that these percentages will be—and should be—very low, as the combined areas of operations for all districts are vast. (Depending on definitions, they may encompass most of the world's oceans.)

Finally, some of our potential metrics for measuring the enforcement of international fishing regulations bear a resemblance to the potential metrics for enforcement of the EEZ. They use percentages of detection and interdiction rates, along with percentages of interdictions leading to successful imposition of penalties, to determine whether Coast Guard actions successfully deter acts of illegal international fishing. Most of these metrics rate as high or medium across the spectrum of validity, reliability, and feasibility, although some of the metrics associ-

[25] USCG, 2004; USCG, 2014a.

ated with accomplishments (as opposed to activities) rate as low for reliability and feasibility. These low-rated metrics require modeling and estimation to determine the frequency of international fishing violations and the impact of Coast Guard actions on that frequency. It is hard to measure the Coast Guard's impact in this context, given the preponderance of natural and human changes that will influence these values; however, better data can inform discussion of and decisions about this mission.

Summary of Findings on Stewardship Missions

As with the safety and security missions, many elements of the logic models for the stewardship missions (ice operations, ATON and WWM, MEP, LMR, and OLE) are not currently being measured (Table 5.6). Again, this reflects the fact that the metrics were developed long before, and independently of, the logic models.

About two-thirds of the stewardship missions' existing metrics were highly feasible; about the same fraction were highly reliable. Only 45 percent of them, though, had high validity; this was an issue across all five missions. Using the potential metrics discussed above can help to increase both the number of elements of the logic models covered and the validity of measurements.

Table 5.6
Numbers of Stewardship Mission Logic Model Elements and Percentages Measured by at Least One Metric

Type of Element	Number	Percentage
Outcomes	8	25
Accomplishments	21	19
Activities	46	30
Total	75	27

CHAPTER SIX

Discussion of Overall Findings

In the preceding three chapters, we discussed existing and potential metrics mission by mission. This chapter captures broad findings on both sets of metrics, reviewing themes and issues that span multiple missions.

Existing Metrics

We observed several key cross-mission issues with respect to sets of existing metrics:

- limited coverage of elements of the logic models
- limited alignment of existing metrics with logic models
- metrics measured in absolute numbers, rather than percentages or rates
- difficulty in defining and measuring denominators
- magnitude limitations.

Limited Coverage of Elements of the Logic Models

Overall, there were three-quarters as many existing metrics (157) total elements for the 11 logic models (213). However, the existing metrics measure fewer than one-third of the elements of the logic models (Table 6.1). Some logic model elements are measured by multiple metrics, but most are not measured at all.

Figure 6.1 presents a more-detailed, mission-by-mission breakdown of this data set. Clearly, there is a great deal of variability. Most elements of most missions' logic models are not captured by existing metrics. This is not surprising, given that the metrics were developed entirely separately and well in advance of the newly created logic models.

Existing metrics tend to measure the most visible, tangible elements of missions, excluding the others. For example, elements dealing with partnerships, intelligence, investigations, and disseminating information tend not to be measured by existing metrics. However, these elements are closely connected with Coast Guard priorities and goals, as enumerated in high-profile Coast Guard documents. For example, the *Coast Guard Western Hemisphere Strategy* is replete with discussion of partnerships, communication, and capacity building.[1] The *Command, Control, Communication, Computers, and Information Technology (C4&IT) Strategic Plan FY13–17* cites improvement of information sharing with partners as its first goal.[2]

1 USCG, 2014d.

2 USCG, 2012b.

Table 6.1
Numbers of Logic Model Elements and Associated Existing Metrics

	A. Number of Each Type of Element	B. Number of Existing Metrics Associated with Each Type of Element	C. Number of Each Type of Element with at Least One Existing Metric	D. Percentage of Each Type of Element with at Least One Existing Metric[a]
Outcomes	20	24	11	55
Accomplishments	50	22	12	24
Activities	144	111	39	27
Total	214	157	62	29

[a] Column C divided by column A.

Figure 6.1
Percentage of Outcomes, Accomplishments, and Activities Measured by at Least One Existing Metric, by Mission

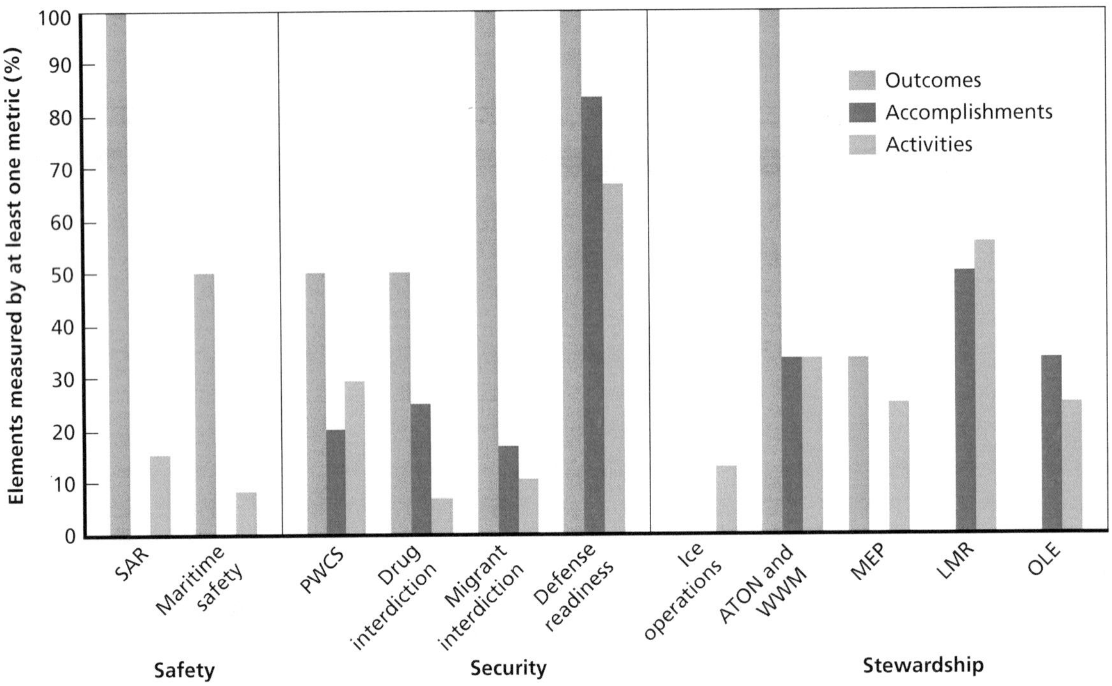

RAND *RR1173-6.1*

Likewise, the *Arctic Strategy* lists broadening partnerships as one of its three strategic objectives.[3] In this context, it is important to be able to measure these elements.

Limited Alignment of Existing Metrics with Logic Models

As was indicated earlier, we evaluated existing metrics in terms of three criteria:

[3] USCG, 2013c.

- **validity**—the extent to which the metric captures the outcome, accomplishment, or activity being assessed
- **reliability**—how consistently measurements can be made
- **feasibility**—how easily the measurement can be made.

Table 6.2 presents numbers and percentages of existing metrics with high values according to these criteria. Most of the metrics have high reliability and feasibility, reflecting the fact that they are relatively objective measures that can be collected without imposing substantial burdens on operators. However, about one-half of them have either medium or low validity with respect to the elements of the logic models. This is not a critique of these metrics' creators—they created these metrics before RAND developed the logic models in late 2014—but an indication that the alignment between existing metrics and the elements of the logic models that were identified from Coast Guard documents and interviews with Coast Guard experts is sometimes limited.

Figure 6.2 shows the percentages of existing metrics that rated as high based on the three criteria, grouped by mission. We see considerable variability among and within missions. One data point that jumps out is that existing defense readiness metrics have no high values for validity. This reflects the fact that existing defense readiness metrics are centered on particular assets, rather than the outcomes, accomplishments, and activities that appear in the logic model: Their measurements are orthogonal to those we are evaluating. Existing defense readiness metrics are typically mapped to multiple elements of the logic model, but with limited validity with respect to any one of them.

Absolute Numbers Versus Percentages or Rates

While some existing metrics are expressed in terms of percentages, a large number are measured in terms of absolute numbers, even though percentages or rates would have provided valuable complementary information. For example, numbers of fatalities, spills, requests met, or interdictions could be divided by one or more appropriate denominators (as they were for other metrics). This would help reveal the extent to which observed trends reflected variables outside the Coast Guard's control, such as increased activity by outside parties.

Conversely, crafting the right denominator and measuring it is also challenging. There are many known unknowns, particularly with respect to low-visibility or clandestine activities, such as the number of recreational boat–miles traveled or the number of FFVs operating in a remote area. Modeling, sampling, and other methods can offer estimates of these denominators, but always with the recognition that there are limitations in terms of reliability, feasibility, or both.

Table 6.2
Percentage of Existing Metrics Evaluated as High

Criteria	Number of Existing Metrics	Percentage of Total (157)
High validity	76	48
High reliability	116	74
High feasibility	129	82
High in all three	53	34

Figure 6.2
Percentage of Existing Metrics Evaluated as High, by Mission

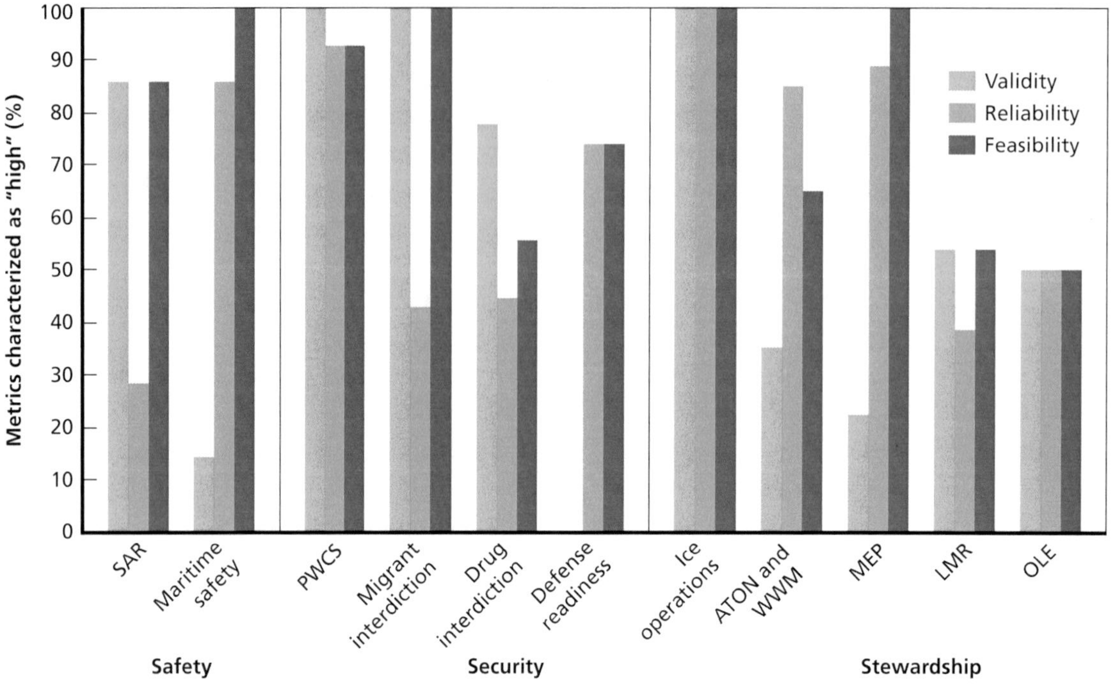

RAND *RR1173-6.2*

Magnitude Limitations

Some existing metrics were limited in that they did not characterize the magnitudes of events or characterized only events of specific magnitudes. Notably, metrics related to spills did not capture the quantity of material spilled. In a different but related vein, existing metrics did not capture SAR cases involving more than ten people or $2 million in property. While this avoids having extreme incidents distort overall figures, it also leaves out important information that could be captured through separate metrics.

Potential Metrics

We recommend implementing only a small fraction of the potential metrics in the appendixes to complement or replace existing metrics. The potential metrics should be viewed as a menu from which the Coast Guard can select, based on a combination of internal priorities and evaluation of the metrics.

To facilitate this selection, we categorized the potential metrics according to how they rated with respect to our three criteria—validity, reliability, and feasibility. As Table 6.3 shows, those that rated as high (H) across all three criteria are likely to be high priorities for the service. Those that had high validity, and at least medium (M) reliability and feasibility, are likely to be useful. Those with medium validity, and at least medium reliability and feasibility, are worth investigating. Finally, those with at least one low (L) value are for potential consideration. We did not discard these metrics from the set we provided to the Coast Guard, because there may be cases in which a metric with a low score represents the best possible way in which to measure an element that may be important to the Coast Guard. Rather, we have noted their deficiencies and are enabling the Coast Guard to decide whether they may be useful despite them.

Table 6.4 provides the numbers of potential metrics that fall into the four categories listed in Table 6.3. About one-quarter of the metrics fall into the high-priority category, with another tenth or so likely to be useful. Collectively, these two categories are roughly as numerous as the existing set of metrics. The table also shows how many of the potential metrics in each category relate to outcomes, accomplishments, and activities.

Table 6.3
Categorization of Potential Metrics According to the Criteria

Validity	Reliability	Feasibility	Category
H	H	H	High priority
H	H or M	H or M	Likely to be useful
M	H or M	H or M	Worth investigating
At least one L	At least one L	At least one L	For possible consideration

Table 6.4
Numbers of Potential Metrics in Different Categories

		Potential Metrics				
	Number of Elements	High Priority	Likely to Be Useful	Worth Investigating	For Possible Consideration	Total
Outcomes	20	18	8	17	16	59
Accomplishments	50	19	12	29	27	87
Activities	144	70	31	155	56	312
Total	214	107	51	201	99	458

CHAPTER SEVEN

Using Metrics to Support Decisionmaking

Metrics provide managers the information they need to make good decisions. The preceding chapters of this report present recommendations for how the Coast Guard should define operational performance and which metrics the Coast Guard could use to plan and evaluate operational performance. The next step in implementing these recommendations is to begin using the metrics within a framework to support decisionmaking. To be useful, a framework for using metrics to support decisionmaking must meet both necessary and sufficient conditions.

Necessary Conditions for Decisionmaking—Validity and Reliability

The necessary conditions for using metrics to support decisionmaking are similar to those for evaluating the metrics presented in Chapters Three, Four, and Five. The metrics must be valid and reliable. When metrics do not represent the organization's objectives, processes, and resources accurately and comprehensively, decisions are less likely to achieve desired results and could lead to unintended, unproductive organizational responses. When metrics do not allow performance to be assessed consistently, leaders can lose confidence in the organization's ability to address the conditions and trends it is confronting.

In developing the recommendations in this report, we sought to achieve these necessary conditions to produce metrics that could support decisionmaking. The logic models for each mission are intended to capture the Coast Guard's understanding of how operations lead to outcomes and strategic objectives. Using validity and reliability to evaluate the proposed metrics provides an initial assessment of whether metrics meet the necessary criteria.

The resulting metrics should, in theory, meet the necessary conditions for supporting decisionmaking. However, during implementation, the proposed logic models must be tested by analyzing performance data, comparing anticipated and achieved results, and adjusting the logic models to reflect new insights into how Coast Guard operations support outcomes and objectives. Such a process is typical of approaches used to manage quality improvement.

The Coast Guard does not currently have a process for using metrics for planning and evaluation that is analogous to the one described in this report and has not formally vetted or adopted the logic models proposed here. It also does not currently collect data for all aspects of our performance logic models. If the Coast Guard decides to use metrics to support operational decisionmaking, ongoing development, vetting, and adaptation of the logic models and associated metrics should be part of a process of continued evaluation and quality improvement for Coast Guard operations.

Sufficient Conditions for Decisionmaking—Breadth and Clarity

Valid and reliable metrics are useful only if they satisfy leadership demands for information that can support decisions. Using a logic model to organize analysis and reporting can support this demand by helping leaders understand how choices affect the full breadth of issues they are responsible for and by clarifying the effects various choices will have on performance, outcomes, and achievement of objectives across the breadth of issues.

Providing Leaders a Holistic, Integrated Analysis of Choices

One fundamental challenge of decisionmaking is the need to prioritize and balance multiple objectives, often against the interests of different organizations, the interests of different geographic regions, or outcomes of different types and consequences. The most desirable choices are those that provide a series of options that have acceptable capability across many objectives and are deficient for only a few.

At the headquarters and area command levels, decisions about how cutters, boats, aircraft, helicopters, billets, and supporting resources are distributed shape levels of operational capability. These choices determine what capability the Coast Guard brings to the nation across its nine districts and 11 missions. It is important to know how resource allocation choices affect the portfolio of Coast Guard capabilities.

The logic models in this report provide an organizing structure for describing plans and performance across the full breadth of Coast Guard missions. Using this structure for analysis and reporting would help leaders understand the implications of their choices across the full breadth of their responsibilities.

Providing Leaders Clarity About the Consequences of Choices

Leaders must also be confident that they understand the consequences of resource allocation decisions. The logic model structure described in this report can provide this kind of clarity in three ways.

First, understanding the consequences of choices requires a detailed understanding of the reasons that capabilities have been or are expected to be as reported. While the logic model structure can provide leaders with breadth in analysis, aggregated, top-level assessments can mask the detailed reasons for one choice being more desirable than another. For example, what does an overall assessment of capabilities for security missions say about the individual missions within that category—PWCS, drug interdiction, migrant interdiction, and defense readiness? In theory, hierarchical decomposition, prescribed by the logic model structure, can be used to provide a detailed analysis for each of the 11 statutory missions, each of their subordinate accomplishments, all the related activities, and all assets. However, in practice, this amount of analysis is both impractical and unnecessary. Details would be necessary only to answer leadership questions and reveal trade-offs within choices. The choices are often isolated in a handful of missions or regions, and detailed analysis can be focused accordingly.

Second, understanding the desirability of an option requires understanding how uncertainty affects outcomes. Uncertainty can arise from many sources, including variability in effectiveness or cost; disagreement about importance of different regions or missions; emergence of uncontrollable events, such as adoption of new technologies or economic booms and busts; or emergence of new adversaries with previously unknown intentions and capabilities. For the Coast Guard, cases of all these uncertainties could influence the desirability of alternative force postures or structures. The following are some examples of factors that create uncertainty for the Coast Guard:

- an adversary
  - a nation-state
  - a trained and sponsored terrorist
  - an inspired individual
  - a vandal
- uncontrollable or unknowable events
  - commodity prices (e.g., oil or natural gas)
  - economic growth in the United States or other countries
  - natural disasters
  - technology advances
  - weather
  - climate change
- variability in performance
  - effectiveness of patrols and surveillance
  - compliance with regulations or guidance.

These cases add dimensionality to analysis, and any or all could be incorporated to the appropriate level of detail in reporting.

Third, understanding evaluations of performance or plans requires understanding the judgments reflected in results. Using the logic model framework to report assessments requires two types of judgment: defining thresholds or goals and selecting methods for aggregating judgments across the logic model hierarchy.

Establishing thresholds and goals is fundamental to defining what constitutes success. In some cases, increased operational capability is always better. In other cases, leadership is satisfied with a specified level of performance, either a minimum acceptable level (i.e., a threshold) or a required level (i.e., a goal). Goals and thresholds can be established in many ways—judging performance against evidenced-based standards, aiming to achieve performance levels established through consensus among SMEs, requiring adherence with best practices, etc. Understanding how success is defined and how acceptable performance levels are set is critical for good decisions. If thresholds and goals are not based on evidence or best practices, the results might be unjustified or arbitrary. Assessments based on such results could lead to choices that introduce undesirable costs or reduce the operational flexibility of competing objectives.

It is also necessary to aggregate assessments at each level of the logic model hierarchy. There are many ways to do this. In some cases, the best way to determine overall performance is to take an average, or weighted average, of the performance of the subordinate assessments. This would, for example, apply if the overall score encompassed assessments of several equally important missions or if performance on measure one might compensate for performance on another, such as an interdiction chain involving detection, response, and resolution rates. In other cases, overall performance is only as good as that of the weakest capability. In some systems, each supporting capability is capable of limiting overall performance. For example, the abilities to process detained illegal migrants and, when necessary, house and return them to their countries of citizenship are all important. And in still other cases, overall performance may be unacceptable if each of its subordinate assessments does not achieve a certain level of performance. For example, the Coast Guard's overall performance with respect to SAR or oil-spill responses could be viewed as a failure unless every district attains a minimum level of performance.

The appropriate approach to use for defining success and aggregating assessments depends on the context and depends on the missions, accomplishments, and activities being assessed; the systems being described; and the extent to which the systems and their functions are understood. Regardless of the approach, for metrics and reporting to be useful to leadership, the judgments incorporated in analysis and evaluation must be transparent.

An Illustrative, Notional Example of Assessment Using the Logic Model Framework

To further explain the concepts described in the preceding sections, this section presents an illustrative analysis of choices the Coast Guard might have to make to adapt its force and fleet posture in response to a reduction in resources. The alternatives and results presented are entirely notional and do not reflect actual assessment of Coast Guard capabilities or performance. However, the approach presented is meant to demonstrate the design and presentation of such an assessment.

Logic models can be viewed in two ways. From an operational perspective, people, assets, and facilities are used to perform tasks and complete core activities. These activities result in accomplishments, which in turn produce mission outcomes in support of strategic outcomes. This bottom-up approach is consistent with a resource-management perspective on performance evaluation.

In contrast, leadership often adopts a strategic perspective, which is built from the same components but is often viewed from the top down. From a strategic perspective, progress on strategic outcomes in part depends on achievement of outcomes, which depend on abilities to achieve accomplishments constituted from activities that rely on assets and resources. This strategic perspective provides a starting point for an assessment that provides the breadth and clarity described in this chapter.

The following example is an illustrative, notional example of an analysis that starts at the top level of the logic model, starting with an assessment of strategic outcomes for the Coast Guard's three primary areas of capability—maritime safety, security, and stewardship—for this case. The example then looks specifically at how different kinds of investment options made in response to a budget reduction might affect these areas. The example then goes on to assess more-detailed levels of the model, and how the lower levels feed into higher-level assessments. At each level, the example looks at a hypothetical baseline case and two notional alternatives, one that targets reductions in resources on specific districts and missions and a second that distributes reductions more evenly, across districts and missions.

Table 7.1 presents an example of an overall, strategic outcome assessment for the example analysis. The overall assessment is that baseline performance is very good (green) for security and stewardship and is acceptable (yellow) for safety. A focused-reduction approach would result in very weak capability for maritime safety (red), an acceptable capability (yellow) for stewardship, and no change for security. In contrast, the evenly distributed alternative reduces performance in all three areas but not as significantly as with a focused reduction. This top-level assessment raises two obvious questions: What constitutes acceptable performance? What is the basis for that judgment? Answering these questions utilizes the hierarchy built into the logic model approach.

To arrive at the overall assessment of maritime stewardship capabilities in Table 7.1, we averaged the assessments for the five missions in Table 7.2. In Table 7.2, the baseline assessment is that the Coast Guard performance is very good (green) for all five missions and that a focused reduction would yield significantly lower but still acceptable (yellow) performance in one, LMR. Evenly distributing the reduction would reduce performance for all five missions, but it would still be good (light green). Performance across missions is consistent in both the baseline and evenly distributed options but varies for the focused-reduction alternative, with the weaker capabilities for the LMR and OLE missions apparently having greater weight.

The use of an averaged aggregation here suggests that, for this decision, leadership is looking to balance capabilities equally across all mission areas. In practice, this may not be the case, and alternative aggregation rules may be more appropriate. However, this illustration demonstrates how such an assessment makes this judgment more transparent.

As before, interpreting the mission-level assessment raises the questions about what constitutes performance and on the basis for assessment. Here again, we can address these questions by digging into a more detail, in this case, by assessing capabilities at the accomplishment level of the logic models. We specifically chose to delve into the LMR mission, which rated yellow in Table 7.2.

The notional aggregated assessment in Table 7.2 was based on how many of the four supporting capabilities—increase awareness, enforce regulations, contribute to conservation, and comply with regulations—the Coast Guard could be expected to perform at acceptable levels (i.e., meet minimum required capabilities). Table 7.3 illustrates assessments of the accomplishment-level capabilities contributing to overall outcomes for the LMR mission. In this notational example, the baseline performance meets required performance for all sup-

Table 7.1
Illustrative, Notional Strategic Outcome Assessment of Coast Guard Expected Capabilities in Response to Potential Resource Reductions

Investment Options	Maritime Safety	Maritime Security	Maritime Stewardship
Baseline			
Focused reduction			
Evenly distributed reduction			

Table 7.2
Illustrative, Notional Mission Outcome Assessment of Coast Guard Expected Capabilities in Response to Potential Resource Reductions

Investment Options	ATON & WWM	Ice Operations	MEP	LMR	OLE
Baseline					
Focused reduction					
Evenly distributed reduction					

Table 7.3
Illustrative, Notional Accomplishment–Level Assessment of Coast Guard Expected Living Marine Resources Capabilities in Response to Potential Resource Reductions

Investment Options	Increase Awareness	Enforce Regulations	Contribute to Conservation	Comply with Regulations
Baseline				
Focused reduction				
Evenly distributed reduction				

porting accomplishments; the focused reduction meets two of four; and the evenly distributed reduction meets three of four.

But how was this performance scored? Once again, the question can be answered through more-detailed analysis. To address this question, we next look specifically at the ability to enforce LMR regulations. We begin by looking at how performance can vary between two regions. Table 7.4 compares operational performance on this metric under the three investment options in two pairs of districts within the Atlantic Area—Districts 1 and 5 in the northeast and Districts 7 and 8 in the south and center. (To reiterate, the data sets and district characterizations are wholly fictional.) In the notional example, with a focused reduction, Districts 7 and 8 could not enforce regulations to the minimum desired operational performance level.

To examine this more closely, Table 7.5 presents a notional analysis of the overall enforcement capability for Districts 7 and 8. In this case, the overall enforcement capability is the product of the percentage of time that the Coast Guard maintains presence in the area of operation; the percentage of time and events that the Coast Guard is able to monitor and detect activity worthy of a response; and the percentage of events for which the Coast Guard is able to successfully respond, interdict, and cite violators. These percentages could be obtained through operational analysis or modeling, and performance might be tracked systematically. In this example, assessment of the focused reduction case suggests that a focused reduction in resources limits activities to maintain presence enough to drop the overall capability below the minimum performance threshold for this activity.

The example presented in this section illustrates how analysis could be framed within the logic model construct and how iterative levels of detailed analysis could be presented to communicate assessments at each level. The discussion showed how judgments at each level feed into one another and can be made transparent. The analysis also demonstrates how an analysis can be expanded to consider cases relevant to leadership, such as differences in performance across regions. Finally, similar drill-downs through the logic model could help leadership understand how alternative force postures and structures can affect the balance of operational performance across other mission areas and performance cases.

Implementing a Framework Using Metrics to Support Decisionmaking

The notional example presented in this section was intended to illustrate how metrics could be used within a framework to support decisionmaking. However, implementing the steps

Table 7.4
Illustrative, Notional Expansion of Analysis to Show District-Level Variation in Assessment of Coast Guard Expected Living Marine Resources Capabilities in Response to Potential Resource Reductions

Investment Options	Districts 1 and 5	Districts 7 and 8
Baseline		
Focused reduction		
Evenly distributed reduction		

Table 7.5
Illustrative, Notional Activity-Level Variation in Assessment of Coast Guard Expected Enforcement of Living Marine Resources Regulations in Response to Potential Resource Reductions

Investment Options	Maintain Presence	Monitor and Detect	Respond, Interdict, and Cite	Overall Enforcement Capability
Baseline	0.95	0.95	0.95	0.86
Focused reduction	0.50	0.95	0.95	0.45
Evenly distributed reduction	0.95	0.95	0.95	0.86

described in this chapter would require concerted and continued effort, most notably in three areas.

First, the approach could be implemented for the problems for which leadership most needs help in making informed decisions. Developing this approach for all 11 mission areas is likely impractical and unnecessary. Instead, implementation could focus on the decisions that appear to be the most important, either because the problems are causing the most concern, because the decisions could affect the most resources, or because the changes in Coast Guard capabilities affect outcomes or objectives the most.

Second, the logic models form the foundation for this decision support framework. However, the Coast Guard has not yet formally accepted the models presented in this report. An initial step toward such acceptance might be to have a broader set of organizations across the Coast Guard and experts familiar with the Coast Guard missions and operations take a closer look at the models for selected missions. The purpose of this review would be to ensure that the logic models for the missions being implemented are not missing important factors or conflict with empirical analysis of the missions.

Finally, refining the logic models should become part of the Coast Guard's efforts for managing operational performance. As mentioned earlier, we drew on documentary and expert sources about the scope and nature of each statutory mission to develop the logic models in this report. However, the completeness and logical coherence of the models has not been tested by gathering data on Coast Guard operations and outcomes. Doing such assessments could offer new insights about the missions that should help refine the logic models and how they are subsequently used to guide planning, analysis, and decisionmaking.

CHAPTER EIGHT

Concluding Remarks

To reiterate, this report was structured around three questions:

- What aspects of operational performance should the Coast Guard aim to measure?
- How well does the Coast Guard measure operational performance?
- How could the Coast Guard measure operational performance better?

To answer the first question, we developed logic models that characterize each of the Coast Guard's 11 statutory missions, based on a bevy of Coast Guard documents and interviews with Coast Guard experts. We answered the second question by evaluating existing metrics derived from the same sources in terms of three criteria: validity (how well they measure an element of the logic model), reliability (how consistently they can be measured), and feasibility (how easily they can be measured). While many existing metrics provide valuable insights on Coast Guard outcomes, accomplishments, or activities, there are also opportunities to improve on existing sets of metrics. We found that 48 percent of existing metrics have high validity, while 74 percent have high reliability, and 82 percent have high feasibility; only 34 percent rate highly on all three criteria. Moreover, existing metrics capture only 29 percent of the elements of the logic models.

The answer to the third question had three components. First, by measuring the 71 percent of the elements of the logic models that existing metrics did not capture, we identified potential metrics that either refined existing metrics by increasing their values with respect to the criteria (particularly validity) or complemented existing metrics. This resulted in a more diversified and refined set of metrics from which the Coast Guard can select. Second, we evaluated these in terms of the same criteria as for existing metrics; this enabled us to organize them into four categories based on the criteria to facilitate Coast Guard decisions about which ones to use. Finally, we developed a framework that Coast Guard leaders could use to review the values of whatever metrics it selects. Given additional prerequisites—such as detailed modeling analyses of relationships among the values of different metrics—the Coast Guard could use such a framework to aid in decisionmaking.

This research provides options for the Coast Guard to consider as part of its overall effort to improve operational planning and evaluation. The Coast Guard can select additional metrics to complement or supersede existing metrics, while also incorporating metrics into a framework to enable senior leadership review. These potential metrics also have relevance beyond the Coast Guard. The newly inaugurated joint task forces can draw on this body of work to consider what metrics are most appropriate for their purposes. The Coast Guard can coordinate with other components of DHS, as well as DoD, NOAA, the Environmental Protection Agency, and other agencies, to align their operational metrics in a way that facilitates communication and decisionmaking.

In addition, the Coast Guard can begin to examine metrics relating to the asset and strategic outcome levels of the logic model, aligning these metrics with the ones discussed in this report. Metrics used to measure mission-support efforts can also be evaluated and aligned. The Coast Guard can also delve more deeply into the proposed framework, evaluating the appropriate values to associate with particular metrics. Finally, the Coast Guard can consider how the values of different metrics relate to one another within a given mission. This is likely to be very challenging; as we have seen, many of the values of particular metrics are influenced by numerous factors beyond the Coast Guard's control. However, understanding these relationships is important in making decisions about resource allocation and for communicating effectively with Congress about the Coast Guard's capacity to achieve particular goals.

APPENDIX A

Search and Rescue

Table A.1
Logic Model for Search and Rescue

Outcomes	Accomplishments	Activities
1. Save and assist lives and property	1.1. Establish communication channels and monitor communications lines	1.1.1. Place and employ communications infrastructure to monitor maritime environment
		1.1.2. Educate maritime community on available resources and how to employ them
		1.1.3. Notify SRU in a timely manner
	1.2. Maintain ready assets and crew for response	1.2.1. Posture assets (e.g., helicopters, boats, cutters) to respond effectively and rapidly
		1.2.2. Ensure that responding crews are able to coordinate with other responding units and agencies when on scene
	1.3. Execute search response operations	1.3.1. Develop search action plan and conduct operational risk evaluation, based on available information and appropriate software
		1.3.2. Deploy the appropriate SRU to execute the response
	1.4. Improve coordination with and capabilities of non–Coast Guard SAR entities	1.4.1. Develop SAR MOUs with strategic partners
		1.4.2. Foster bilateral, regional, interagency, international, private-sector, and other external SAR engagement
		1.4.3. Assist in the development of SAR policy, requirements, programs, and plans
		1.4.4. Establish domestic and international partnerships to save lives in the most effective manner
		1.4.5. Coordinate AMVER
		1.4.6. Respond effectively to SURPIC requests

Table A.2
Existing Metrics for Search and Rescue

Outcome, Accomplishment, or Activity	Existing Metric	Source Document(s)	Comments on the Metric	Validity	Reliability	Feasibility	Comments on Ratings
1. Save and assist lives and property	Percentage of people in imminent danger saved in the maritime environment	COMDTINST M16130.2F; Atlantic Area OPD (LANT-OPD)		H	M	H	Imminence of danger may be subjective; does not include SAR cases involving 11 or more lives
	Percentage of all mariners in distress saved after Coast Guard notification	COMDTINST M16130.2F; LANT-OPD		H	M	H	Distress may be subjective; does not include SAR cases involving 11 or more lives
	Percentage of property in danger of loss saved	COMDTINST M16130.2F; LANT-OPD		H	M	H	Does not include property losses valued at more than $2 million
	Property (in millions of dollars) at risk	COMDTINST M16130.2F; LANT-OPD		H	M	H	Does not include property valued at more than $2 million
1.1.1. Place and employ communications infrastructure to monitor maritime environment	Percentage of continuous coverage for reception of a 1-W signal of a 1-m antenna, out to 20 nmi from shore, as part of the National Distress and Response System	COMDTINST M16130.2F	Excludes select parts of Alaska	M	M	L	
1.2.1. Posture assets (e.g., helicopters, boats, cutters) to respond effectively and rapidly	Percentage of incidents in which suitable SAR resource ready to proceed within 30 minutes of notification of distress	COMDTINST M5000.3B; LANT-OPD	In MISLE but not tracked collectively across all cases Does not include transit time	H	H	H	
	Percentage of incidents in which Coast Guard SRU arrives on scene within two hours of notification	COMDTINST M16130.2F; LANT-OPD	The two hours includes the 30 minutes for time to proceed, but does not include transit time	H	H	H	

Table A.3
Potential Metrics for Search and Rescue

Associated Activity, Accomplishment, or Outcome	Potential Metric	Comments on the Metric	Validity	Reliability	Feasibility	Comments on Ratings
1. Save and assist lives and property	Percentage of mass-casualty incidents in an FY in which the circumstances exceeded Coast Guard organic capabilities		H	H	H	
	Percentage of mass-casualty (>10 people) incidents in which all individuals were saved		H	H	H	
	Percentage of individuals saved in mass-casualty incidents		H	H	H	
	Percentage of people in imminent danger saved in the maritime environment		H	H	H	
	Percentage of all people in distress saved after Coast Guard notification		H	H	H	
	Number of people assisted (non–life threatening situation)		H	H	H	
	Number of search operations that had to be suspended		H	H	H	
	Percentage of property in danger of loss saved		H	H	H	
	Property (in millions of dollars) at risk		H	H	H	
	Number of high-property-value (>$2 million) incidents and percentage saved		H	H	H	
	Types of vessels assisted		H	H	H	

Table A.3—Continued

Associated Activity, Accomplishment, or Outcome	Potential Metric	Comments on the Metric	Validity	Reliability	Feasibility	Comments on Ratings
1.1. Establish communication channels and monitor communications lines	Aggregate metric: average of metrics associated with subordinate activities		M	M	H	
1.1.1. Place and employ communications infrastructure to monitor maritime environment	Percentage of SAR cases in which communications infrastructure shortfalls were determined to have reduced the effectiveness of response		H	M	M	
	Percentage of SAR cases in which Rescue 21 towers' triangulation of communications sources provided directional data		M	M	M	
1.1.2. Educate maritime community on available resources and how to employ them	Percentage of goal audience reached per FY	This would require surveys	H	M	L	
1.1.3. Notify SRU in a timely manner	Percentage of instances in which action is taken within 5 minutes		H	H	H	
1.2. Maintain ready assets and crew for response	Aggregate metric: average of metrics associated with subordinate activities		M	M	H	
1.2.1. Posture assets (e.g., helicopters, boats, cutters) to respond effectively and rapidly	Percentage of cases in which suitable SAR resource ready to proceed within 30 minutes of notification of distress		M	H	H	
	Percentage of cases in which Coast Guard SRU arrives on scene within two hours of notification or within logistically feasible time line	There are locations (e.g., Guam, remote portions of Alaska) that are logistically impossible to reach within two hours of the nearest Coast Guard asset	M	H	H	

Table A.3—Continued

Associated Activity, Accomplishment, or Outcome	Potential Metric	Comments on the Metric	Validity	Reliability	Feasibility	Comments on Ratings
	Percentage of cases an SRU is pulled from one operational mission and tasked to a higher priority		M	H	H	
1.2.2. Ensure that responding crews are able to coordinate with other responding units and agencies when on scene	Percentage of request SRUs reporting any on-scene communications failures		H	M	M	
1.3. Execute search response operations	Percentage of SAR asset-hours and personnel-hours spent on incidents in which no people or property were at risk		H	H	H	
	Percentage of SAR asset-hours and personnel-hours spent on false alerts and hoaxes		H	H	H	
	The number of cases or responses where SRUs respond in either red or amber risk category (corresponding to higher risk)		H	H	H	
	Average time spent searching for a person in distress		H	H	H	
1.3.1. Develop search action plan and conduct operational risk evaluation, based on available information and appropriate software	Percentage of cases in which the SAR object is located within the search area		M	H	H	
1.3.2. Deploy the appropriate SRU to execute the response	Percentage of cases in which insufficient or no assets were available to respond to SAR cases		M	M	M	

Table A.3—Continued

Associated Activity, Accomplishment, or Outcome	Potential Metric	Comments on the Metric	Validity	Reliability	Feasibility	Comments on Ratings
1.4. Improve coordination with and capabilities of non–Coast Guard SAR entities	Number of incidents in which the capabilities of a partner were required or requested and were unable to be fulfilled		M	H	H	
	Number of SAR training events held with partners each FY		M	H	H	
	Identified areas in which peer organizations have greater capabilities than the Coast Guard	This probably requires querying SMEs	M	L	M	
1.4.1. Develop SAR MOU with strategic partners	Number of MOUs or MOAs divided by number desired		H	H	H	
1.4.2. Foster bilateral, regional, interagency, international, private-sector, and other external SAR engagement	Number of bilateral engagements the Coast Guard participated in to promote SAR		H	H	H	
1.4.3. Assist in the development of SAR policy, requirements, programs, and plans	Number of incidents in which circumstances existed in which the current policy or requirements were not applicable and therefore needed to be examined for potential revision		M	H	M	
1.4.4. Establish domestic and international partnerships to save lives in the most effective manner	Number of SAR events in which the Coast Guard participated with an international partner		H	H	H	
1.4.5. Coordinate AMVER	Percentage of SAR events successfully resolved via an AMVER participant		H	H	H	
1.4.6. Respond effectively to SURPIC requests	Percentage of SURPIC requests fulfilled		H	H	H	

Marine Safety

Table B.1
Logic Model for Marine Safety

Outcomes	Accomplishments	Activities
1. Reduce the frequency and extent of injuries, fatalities, property damage, and environmental damage due to safety incidents	1.1. Increase compliance by mariners, marine facilities, and vessels	1.1.1. Monitor, inspect, examine, and investigate facilities, vessels, and contents (such as containers, bulk goods, and ballast water)
		1.1.2. Cite non–incident related safety worklist items from facility and vessel inspections and note accordingly in MISLE casework narrative
	1.2. Increase stakeholder knowledge base and documentation of it	1.2.1. Provide consultation to industry advisory bodies and conduct maritime sector outreach
		1.2.2. Administer licensing and credentialing program for professional mariners and vessel documentation and decal programs
		1.2.3. Provide regulatory, safety, and compliance guidance, as well as information on best practices, to all stakeholders
		1.2.4. Monitor and report North Atlantic iceberg conditions using fixed-wing aircraft and reports from ships as part of the International Ice Patrol
2. Investigate and document safety incidents	2.1. Investigate incident	2.1.1. Document incident and collect evidence
		2.1.2. Determine incident causal factors
		2.1.3. Identify responsible parties
		2.1.4. Assist interagency, state, and local investigations
		2.1.5. Disseminate investigation findings for use in improved regulation to field and industry
	2.2. Prepare for legal proceedings	2.2.1. Issue appropriate citations
		2.2.2. Prepare case package for prosecution

Table B.2
Existing Metrics for Marine Safety

Outcome, Accomplishment, or Activity	Existing Metric	Source Document(s)	Comments on the Metric	Validity	Reliability	Feasibility	Comments on Ratings
1. Reduce the frequency and extent of injuries, fatalities, property damage, and environmental damage due to safety incidents	Commercial fishing vessel (CFV) fatalities	OPAR	MISLE	H	H	H	
	CFV losses	OPAR	MISLE	H	H	H	
	Commercial mariner deaths and injuries	OPAR	MISLE	H	H	H	
	Commercial passenger deaths and injuries	OPAR	MISLE	H	H	H	
	Number of oil spills over 100 gallons	Marine Safety Mission Performance Plan FY11–16	Target FYs 2015–2016: ≤131	M	M	H	Number of spills is of interest, but sizes of spills and degree of hazard associated with nonoil chemicals may be more important; medium feasibility for numbers of spills (without reference to size) because small spills may not be documented
	Five-year average number of oil spills over 100 gallons per 100 million short tons shipped	Maritime Prevention Prog Perf Plan FY14–FY19, August 2014; Michel, 2014	Target: ≤11.1 (FY 2015)	M	M	H	
	Number of chemical discharge incidents	Marine Safety Mission Perf Plan FY11–16	Targets: FY 2015 ≤ 28; FY 2016 ≤ 27	M	M	H	
	Average number of chemical discharge incidents	Michel, 2014	Management measure, ≤15.8	M	M	H	

Table B.2—Continued

Outcome, Accomplishment, or Activity	Existing Metric	Source Document(s)	Comments on the Metric	Validity	Reliability	Feasibility	Comments on Ratings
	Five-year average number of chemical discharge incidents per 100 million short tons shipped	Maritime Prevention Prog Perf Plan FY14–FY19, Aug 2014	Targets: ≤15.9 9 (FY14); ≤15.8 (FY 2015)	M	M	H	
1.1.1. Monitor, inspect, examine, and investigate facilities, vessels, and contents (such as containers, bulk goods, and ballast water)	Number of random containers inspected	Frank Albero, "Info Paper: Review of LANTAREA Operational Planning Direction (OPD) Performance Measures," memorandum, Boston, Mass.: U.S. Coast Guard First District, April 8, 2014, Not available to the general public		M	H	H	
	Total number of containers inspected	Albero, 2014		M	H	H	
	Percentage of containers randomly inspected	Maritime Prevention Program Performance Plan FY14–FY19, Aug 2014	Target: 10% of total inspections; conducted on general cargo shipment to find undeclared hazardous materials	M	H	H	
	Uninspected towing vessel dockside exams	Michel, 2014; CGBI	Data sources: MISLE, CGBI measure ID: 22047	M	H	H	
	Percentage of uninspected towing vessel fleet boardings	Michel, 2014	Data source: MISLE	M	H	H	
	Percentage of Maritime Labor Convention trade vessels inspected	Michel, 2014	Data source: MISLE	M	H	H	
	CFV safety dockside exams	Michel, 2014; CGBI	Data sources: MISLE; CGBI measure ID: 19121	M	H	H	

Table B.2—Continued

Outcome, Accomplishment, or Activity	Existing Metric	Source Document(s)	Comments on the Metric	Validity	Reliability	Feasibility	Comments on Ratings
	Underway boardings of CFVs	Michel, 2014; CGBI	Data sources: MISLE; CGBI measure ID: 20793	M	H	H	
	CFV safety decals and certificates of compliance issued	Michel, 2014	Data source: MISLE	M	H	H	
	Number of port state control exams	Michel, 2014; CGBI	Data sources: MISLE; CGBI measure ID: 19115	M	H	H	
	Number of International Ship & Port Security (ISPS) exams, validity actions, and detentions	Michel, 2014; CGBI	Data sources: MISLE; CGBI measure IDs: 21376, 21537, 21386, 21377, 21536	M	H	H	
	Percentage of facility safety inspections at Maritime Transportation Security Act of 2002 (MTSA)–regulated waterfront facilities	Michel, 2014; CGBI	Data sources: MISLE; Target: 100%; CGBI measure ID: 22239	M	H	H	
	Number of facility safety inspections at non-MTSA (33 Code of Federal Regulations–regulated) waterfront facilities	Michel, 2014; CGBI	Data sources: MISLE; CGBI measure ID: 22240	M	H	H	
	Percentage of MARPOL waste reception facility reports of inadequacy investigated	Michel, 2014; CGBI; Maritime Prevention Prog Perf Plan FY14–FY19; Aug 2014; Albero, 2014	Data sources: MISLE; Target: 100%; CGBI measure ID: 22122	M	H	H	
	Percentage of remedial action issues completed	Michel, 2014; Brown, 2013	Data sources: Contingency Preparedness System; Target: 80% resolved (75% target source: Brown, 2013)	M	H	H	

Table B.2—Continued

Outcome, Accomplishment, or Activity	Existing Metric	Source Document(s)	Comments on the Metric	Validity	Reliability	Feasibility	Comments on Ratings
	Percentage of remedial action issues outstanding	Michel, 2014	Data sources: Contingency Preparedness System; Target: 85% resolved (total number resolved after 18 or more months)	M	H	H	
	Number of transfer monitors	Michel, 2014; CGBI	Data sources: MISLE; CGBI measure ID: 21575	M	H	H	
	Ballast water examinations	CGBI	CGBI measure ID: 19628; count of all ballast water examination subactivities that have been conducted	M	H	H	
	Response boat–small boardings	OPAR	MISLE	H	H	H	
	Uninspected passenger vessel dockside inspections			M	H	H	
	U.S. inspected passenger vessel fleet inspections			M	H	H	
	U.S. inspected deep draft vessel inspections			M	H	H	
	U.S. inspected barge fleet inspections			M	H	H	
	Offshore supply vessel fleet inspections			M	H	H	
	Outer continental shelf inspections			M	H	H	
	Life raft inspections			M	H	H	

Table B.3
Potential Metrics for Marine Safety

Associated Activity, Accomplishment, or Outcome	Potential Metric	Comments on the Metric	Validity	Reliability	Feasibility	Comments on Ratings
1. Reduce the frequency and extent of injuries, fatalities, property damage, and environmental damage due to safety incidents	Fatalities per 100,000 licensed mariners		M	H	H	
	Serious injuries per 100,000 licensed mariners		M	H	H	
	Passenger fatalities per 100,000 passenger-trips		M	M	M	
	Passenger serious injuries per 100,000 passenger-trips		M	M	M	
	Recreational boater fatalities per 100,000 person-trips on boats		M	M	M	
	Recreational boater serious injuries per 100,000 person-trips on boats		M	M	M	
	Property damage in millions of dollars per billion dollars of facilities and vessels in the		M	L	L	
	Environmental damage, measured in terms of dollar impact, divided by value of maritime activity (commerce handled, passenger travel, fish, recreational benefits)		M	L	L	
1.1. Increase compliance by mariners, marine facilities, and vessels	Percentage compliance rate by mariners, marine facilities, and vessels, by category of type of compliance or regulation	This needs to be estimated or extrapolated from samples	H	M	L	

Table B.3—Continued

Associated Activity, Accomplishment, or Outcome	Potential Metric	Comments on the Metric	Validity	Reliability	Feasibility	Comments on Ratings
1.1.1. Monitor, inspect, examine, and investigate facilities, vessels, and contents (such as containers, bulk goods, and ballast water)	Number of inspections, examinations, and investigations, by type		H	M	H	
	Percentage of relevant entities inspected, examined, and/or investigated		H	M	M	
1.1.2. Cite non–incident-related safety worklist items from facility and vessel inspections and note accordingly in MISLE casework narrative	Number of non–incident-related safety worklist items divided by number of entities inspected, examined, or investigated		M	H	H	
	Percentage of items found requiring citation that were cited		M	H	H	
1.2. Increase stakeholder knowledge base and documentation of it	Estimated percentage of relevant audiences learning key information	This would require surveys that included quizzes	M	M	L	
1.2.1. Provide consultation to industry advisory bodies and conduct maritime sector outreach	Number of meetings with industry advisory bodies		L	H	H	
1.2.2. Administer licensing and credentialing program for professional mariners and vessel documentation and decal programs	Percentage of professional mariners who are licensed	The numerators are relatively easy to get, but the denominators are more challenging and less accurate	M	M	M	
	Percentage of professional mariners who are credentialed		M	M	M	
	Percentage of vessels with decals		M	M	M	

Table B.3—Continued

Associated Activity, Accomplishment, or Outcome	Potential Metric	Comments on the Metric	Validity	Reliability	Feasibility	Comments on Ratings
1.2.3. Provide regulatory, safety, and compliance guidance, as well as information on best practices, to all stakeholders	Forms of communication being used		M	H	H	
	Estimated percentages of relevant audiences being reached	Accuracy would require surveys	M	M	L	
	Recreational boating safety boardings divided by total number of registered recreational boats		M	H	H	
	Percentage of boarded recreational boats with safety discrepancies		M	H	H	
1.2.4. Reduce the risk of collision with icebergs as part of the International Ice Patrol	Number of vessel collisions with icebergs in a given season in the North Atlantic		H	H	H	
	Percentage of vessels transiting the North Atlantic in a given season that collide with icebergs	This percentage will be very small; it could also be presented as number of collisions per 100,000 transits	H	H	H	
	Number of iceberg warning products provided in a given season		M	H	H	
	Percentage of iceberg warning products provided within a given number of hours of receipt of information from aircraft and ships		M	H	H	
	Frequency with which given areas are monitored for icebergs		M	H	H	
2. Investigate and document safety incidents	Percentage of cases in which the Coast Guard completes an investigation		L	H	H	

Table B.3—Continued

Associated Activity, Accomplishment, or Outcome	Potential Metric	Comments on the Metric	Validity	Reliability	Feasibility	Comments on Ratings
	Estimated impact on compliance of prosecution	This would require modeling and surveys	M	L	L	
2.1. Investigate incident	Percentage of cases in which the Coast Guard conducts an investigation		L	H	H	
2.1.1. Document incident and collect evidence	Percentage of incidents resulting in a report		M	H	H	
2.1.2. Determine incident causal factors	Percentage of incidents in which causal factors are determined		H	H	H	
2.1.3. Identify responsible parties	Percentage of incidents in which responsible parties are identified		H	H	H	
2.1.4. Assist interagency, state, and local investigations	Percentage of incidents in which the Coast Guard assists other entities in their investigations	There are degrees of assistance, but the focus is on a binary distinction—whether the Coast Guard helped at least one other agency investigate any given incident	M	H	H	
2.1.5. Disseminate investigation findings for use in improved regulation to field and industry	Estimated percentage of relevant audiences reached	To achieve accuracy, this would require surveys	M	M	L	
2.2. Prepare for legal proceedings	Percentage of successful prosecutions		L	H	M	
2.2.1. Issue appropriate citations	Number of instances in which citations are issued divided by number in which they should be		M	M	H	
2.2.2. Prepare case package for prosecution	Percentage of case packages completed		M	M	H	

APPENDIX C

Ports, Waterways, and Coastal Security

Table C.1
Logic Model for Ports, Waterways, and Coastal Security

Outcomes	Accomplishments	Activities
1. Enhance preparedness for and prevention of maritime terrorist attacks	1.1. Ensure vessel and facility compliance with maritime security laws and regulations	1.1.1. Ensure plans are in place, and review to ensure adequacy
		1.1.2. Conduct inspections and boardings to ensure plans are followed
		1.1.3. Enforce TWIC
		1.1.4. Conduct external patrols to assess security measures
	1.2. Collect intelligence in collaboration with other agencies and entities	1.2.1. Collect intelligence
		1.2.2. Contribute to collaborative bodies and organizations to analyze, integrate, and disseminate intelligence
		1.2.3. Share relevant intelligence information with other stakeholders as appropriate
	1.3. Ensure PWCS preparedness	1.3.1. Conduct Coast Guard exercises
		1.3.2. Conduct exercises with external entities
		1.3.3. Conduct outreach with the public
		1.3.4. Engage with the other members of the AMSC
2. Deter potential maritime terrorist attacks; counter and respond to actual maritime terrorist attacks	2.1. Maintain presence near critical or vulnerable targets	2.1.1. Conduct waterborne, air, or shoreside patrols and surveillance around high-risk infrastructure or events
		2.1.2. Escort vessels
		2.1.3. Conduct random and targeted security boardings of vessels
		2.1.4. Enforce fixed security zones
	2.2. Counter terrorist attacks when cued by intelligence or events	2.2.1. Surge to respond to security threats when alerted
		2.2.2. Provide TPS

Table C.2
Existing Metrics for Ports, Waterways, and Coastal Security

Outcome, Accomplishment, or Activity	Existing Metric	Source Document(s)	Comments on the Metric	Validity	Reliability	Feasibility	Comments on Ratings
1.1. Ensure vessel, organizational, and facility compliance with maritime security laws and regulations	Percentage security compliance rate for high-risk maritime facilities	2014 Third Quarter Performance Review, p. 7		H	H	H	
	Percentage security compliance rate for high-risk vessels			H	H	H	
	Percentage security compliance rate for high-risk offshore facilities			H	H	H	
1.1.1. Ensure plans are in place and review to ensure adequacy	Percentage of required area maritime security plan exercises conducted	Michel, 2014, p. 33		H	H	H	
1.1.2. Conduct inspections and boardings to ensure plans are followed	Percentage of announced deepwater port facility security inspections	Servidio, 2013b, p. 75		H	H	H	
	Percentage of announced and unannounced facility and security inspections for MTSA facilities and non-MTSA facilities	Servidio, 2013b, p. 75; OPAR		H	H	H	
	Percentage of high-interest vessels boarded	Michel, 2014, p. 10	Data source: MSRO Manual Appendix	H	H	H	
	Percentage compliance with USN transit protection requirements	Michel, 2014, p. 10	Data source: Coast Guard–USN MOA	H	H	H	
	Percentage of small vessel security boardings	2014 Third Quarter Performance Review, p. 14		H	H	H	

Table C.2—Continued

Outcome, Accomplishment, or Activity	Existing Metric	Source Document(s)	Comments on the Metric	Validity	Reliability	Feasibility	Comments on Ratings
	Percentage of random target container inspections (on general cargo shipments, to find undeclared hazardous materials)	Michel, 2014, p. 30		H	H	H	
1.1.3. Enforce TWIC	Percentage of facility MTSA compliance exams and MTSA verifications	Michel, 2014, p. 33		H	H	H	
	TWIC checks	Michel, 2014, p. 33		H	H	H	Denominator and actual percentage based on maritime security level
	TWIC compliance rate	OPAR		H	M	M	
2. Deter potential maritime terrorist attacks; counter and respond to actual maritime terrorist attacks	Maritime security risk reduction: consequence management, terrorist transfer, weapons of mass destruction transfer	2014 Third Quarter Performance Review, p. 14		H	L	L	
2.1.1. Conduct waterborne, air, or shoreside patrols and surveillance around high-risk infrastructure or events	Percentage of positive control measures enacted for high-interest vessels	MSRO Scorecard		H	H	H	
	Percentage of positive control measures enacted for non–high-interest vessels	MSRO Scorecard		H	H	H	
	Percentage of positive control measures enacted for certain dangerous cargo	MSRO Scorecard		H	H	H	

Table C.2—Continued

Outcome, Accomplishment, or Activity	Existing Metric	Source Document(s)	Comments on the Metric	Validity	Reliability	Feasibility	Comments on Ratings
	Percentage of patrols around fixed security zones	MSRO Scorecard		H	H	H	Denominator determined by the Maritime Security Risk Analysis Model (MSRAM)
	Percentage of patrols around permanent marine critical infrastructure and key resources	MSRO Scorecard		H	H	H	Denominator determined by MSRAM
	Percentage of patrols around conditional marine critical infrastructure and key resources	MSRO Scorecard		H	H	H	Denominator determined by MSRAM
2.1.2. Escort vessels	Percentage of high- value unit vessels escorted	MSRO Scorecard		H	H	H	
	Percentage of vessels carrying certain dangerous cargo escorted	MSRO Scorecard; Michel, 2014, p. 10	Data source: MSRO Manual Appendix	H	H	H	
	Percentage of non–high-interest vessels boarded	MSRO Scorecard		H	H	H	
	Percentage of high-capacity passenger vessels escorted	Michel, 2014, p. 10	Data source: MSRO Manual Appendix	H	H	H	
	Percentage of sealift vessels escorted	MSRO Scorecard		H	H	H	
	Percentage of security boardings conducted for non–high-interest vessels (military outload)	MSRO Scorecard		H	H	H	
	Percentage of positive control measures enacted for non–high-interest vessels (military outload)	MSRO Scorecard		H	H	H	

Table C.3
Potential Metrics for Ports, Waterways, and Coastal Security

Associated Activity, Accomplishment, or Outcome	Potential Metric	Comments on the Metric	Validity	Reliability	Feasibility	Comments on Ratings
1. Enhance preparedness for and prevention of maritime terrorist attacks	Estimated percentage risk reduction associated with intelligence preparation and enhanced compliance	Extraordinarily difficult to calculate accurately; however, risk modeling and SME inputs can enable it to be calculated to some degree of approximation	H	L	L	
1.1. Ensure vessel, organizational, and facility compliance with maritime security laws and regulations	None needed—covered by existing metrics					
1.1.1. Ensure plans are in place and review to ensure adequacy	Percentage of permanent and nonpermanent marine critical infrastructure plans reviewed		H	H	H	
	Percentage of compliant plans		H	M	H	
1.1.2. Conduct inspections and boardings to ensure plans are followed	None needed—covered by existing metrics					
1.1.3. Enforce TWIC	Percentage of workers checked		H	H	M	
1.1.4. Conduct external patrols to assess security measures	Percentage of facilities with significant security deficiencies observed by maritime and shoreside patrols		M	M	M	
1.2. Collect intelligence in collaboration with other agencies and entities	External assessment of the quality of Coast Guard PWCS-related intelligence contributions on a 0–10 scale		M	M	M	
1.2.1. Collect intelligence	Number of full-time Coast Guard personnel focusing on PWCS-related intelligence collection		L	H	H	

Table C.3—Continued

Associated Activity, Accomplishment, or Outcome	Potential Metric	Comments on the Metric	Validity	Reliability	Feasibility	Comments on Ratings
	External assessment of the quality of Coast Guard PWCS-related intelligence collection on a 0–10 scale	Would require a survey	M	M	M	
	Number of America's Waterway Watch reports		L	H	H	
	Number of America's Waterway Watch investigations		L	H	H	
	Number of small-vessel security cases and investigations		L	H	H	
1.2.2. Contribute to collaborative bodies and organizations to analyze, integrate, and disseminate intelligence	Percentage of requests for participation in meetings and conferences supported		M	H	H	
	Number of Coast Guard personnel contributing to collaborative bodies and organizations addressing PWCS		M	H	H	
	Number of full-time Coast Guard personnel working at bodies and organizations addressing PWCS		M	H	H	
1.2.3. Share relevant intelligence information with other stakeholders as appropriate	Percentage of relevant external audience that knows about Coast Guard intelligence reports	Would require a survey of the defined relevant external audiences—e.g., CBP intelligence personnel	H	M	L	
	Percentage of relevant external audience that has read Coast Guard intelligence reports	Would require a survey of the defined relevant external audiences—e.g., CBP intelligence personnel	H	M	L	
	Percentage of requests for information answered		H	M	H	

Table C.3—Continued

Associated Activity, Accomplishment, or Outcome	Potential Metric	Comments on the Metric	Validity	Reliability	Feasibility	Comments on Ratings
	Percentage of nonrequests for information sharing opportunities supported		M	M	H	
1.3. Ensure PWCS preparedness	Estimated risk reduction due to exercises and outreach	Extraordinarily difficult to calculate accurately; however, risk modeling can enable it to be calculated to some degree of approximation	H	L	L	
1.3.1. Conduct Coast Guard exercises	Number of Coast Guard PWCS exercises		M	H	H	
	Number of Coast Guard small-vessel security exercises		M	H	H	
1.3.2. Conduct exercises with external entities	Number of exercises involving other partners		M	H	H	
	Number of small-vessel security exercises involving other agencies		M	H	H	
	Number of small-vessel security exercise participants		M	H	H	
	Percentage of desired participants in small-vessel security exercises		M	M	M	
1.3.3. Conduct outreach	Number of small-vessel security awareness, training, and education programs		M	H	H	
	Number of participants in small-vessel security awareness, training, and education programs		M	H	H	

Table C.3—Continued

Associated Activity, Accomplishment, or Outcome	Potential Metric	Comments on the Metric	Validity	Reliability	Feasibility	Comments on Ratings
	Percentage of desired audience participating in small-vessel security awareness, training, and education programs		M	M	M	
	Percentage of participants retaining desired knowledge	Would require quizzes	M	M	L	
	Number of joint efforts with international organizations as a percentage of joint efforts sought		M	M	H	
	Number of improvements in U.S. trading-partner ports beyond ISPS code as a percentage of improvements sought		M	M	H	
1.3.4. Engage with the other members of the AMSC	Number of AMSC meetings per year		M	H	H	
	Number of meetings with stakeholders belonging to the AMSC per year		M	M	M	
2. Deter potential maritime terrorist attacks; counter and respond to actual maritime terrorist attacks	Number of injuries due to maritime terrorist attacks	Number depends on many factors besides the Coast Guard's actions, notably terrorists' interests and capabilities	L	H	H	
	Number of fatalities due to maritime terrorist attacks		L	H	H	
	Extent of property damage, in millions of dollars, due to maritime terrorist attacks		L	H	H	

Table C.3—Continued

Associated Activity, Accomplishment, or Outcome	Potential Metric	Comments on the Metric	Validity	Reliability	Feasibility	Comments on Ratings
	Percentage of maritime terrorist attacks that achieve their full intended impact	Will usually be incalculable because the percentages involve dividing zero by zero and/or knowing the terrorists' intentions; performance in exercises could be used as a proxy	M	L	L	
	Number of successful attacks per year	Will usually be zero; however, number speaks to the effectiveness of both deterrence and response	H	H	H	
	Number of known attempted attacks per year		H	H	H	
	Number of known planned attacks per year		H	M	M	This may have some ambiguity—e.g., whether multiple planned attacks are really elements of a single attack and whether the plans rose to a meaningful level or were merely the speculations of disgruntled individuals
	Percentage of maritime terrorist attacks that have a reduced impact due to Coast Guard actions		M	M	H	
	Percentage of maritime terrorist attacks that have a reduced impact due to all agencies' actions		M	M	H	
	Percentage of maritime terrorist attacks prevented from having an impact (injuries, fatalities, major property damage) by Coast Guard actions		M	M	H	

Table C.3—Continued

Associated Activity, Accomplishment, or Outcome	Potential Metric	Comments on the Metric	Validity	Reliability	Feasibility	Comments on Ratings
	Percentage of maritime terrorist attacks prevented from having an impact (injuries, fatalities, major property damage) due to all agencies' actions		M	M	H	
2.1. Maintain presence near critical or vulnerable targets	Percentage of time Coast Guard forces are near critical or vulnerable targets	Given numerous potential targets, this requires either averaging evenly over them all (i.e., one liquefied natural gas tanker is equivalent to the Statue of Liberty) or weighting them to reflect risks; note that a lack of adversary ability to anticipate may not always be a good thing (may reduce deterrence in some cases)	M	L	M	
	Percentage of time Coast Guard forces are near critical or vulnerable targets and the Coast Guard's presence would be difficult for an adversary to anticipate	Likely requires wargaming or live exercises involving a red cell	M	L	L	
2.1.1. Conduct waterborne, air, or shoreside patrols and surveillance	Percentage of time the Coast Guard is patrolling high-risk infrastructure	Given numerous pieces of infrastructure, this requires either averaging evenly over all of them (i.e., one liquefied natural gas tanker is equivalent to the Statue of Liberty) or weighting them to reflect importance, vulnerability, and other attributes related to risk	M	L	M	

Table C.3—Continued

Associated Activity, Accomplishment, or Outcome	Potential Metric	Comments on the Metric	Validity	Reliability	Feasibility	Comments on Ratings
	Percentage of time the Coast Guard is paroling high-risk infrastructure or events and that it would be difficult for an adversary to anticipate	Likely requires wargaming or live exercises involving a red cell; note that a lack of adversary ability to anticipate may not always be a good thing (may reduce deterrence in some cases)	M	L	L	
2.1.2. Escort vessels	Number of vessels escorted		M	H	H	
	Percentage of workloads for escort met		H	H	H	
	Percentage of all vessels escorted		M	H	M	
2.1.3. Conduct random and targeted security boardings of vessels	Percentage of vessels boarded in a given area and time frame		M	M	M	
	Percentage of vessels targeted for security boardings that are actually boarded		M	H	H	
2.1.4. Enforce fixed security zones	Number of response assets per accessible boundaries or area of fixed security zone	Can be parsed by type of response asset (e.g., small boat, helicopter); the denominator can be either the boundaries or the area, depending on what best reflects tactical capabilities	M	H	H	
2.2. Counter terrorist attacks when cued by intelligence or events	Percentage of maritime terrorist attacks in which the Coast Guard is able to respond	This will usually be incalculable, since it involves dividing zero by zero Performance in exercises could be used as a proxy	M	M	H	
2.2.1. Surge to respond to security threats when alerted	Percentage of requests for surge support met		H	H	H	

Table C.3—Continued

Associated Activity, Accomplishment, or Outcome	Potential Metric	Comments on the Metric	Validity	Reliability	Feasibility	Comments on Ratings
	Percentage of events requiring surge support supported		H	H	H	
2.2.2. Provide TPS	Percentage of TPS requests met		H	H	H	

APPENDIX D

Drug Interdiction

Table D.1
Logic Model for Drug Interdiction

Outcomes	Accomplishments	Activities
1. Reduce the maritime flow of illegal drugs	1.1. Deter, divert, detain, and disrupt maritime drug-smuggling flows	1.1.1. Detect and monitor
		1.1.2. Intercept
		1.1.3. Interdict
		1.1.4. Board and apprehend
		1.1.5. Achieve visible or perceived presence
	1.2. Support and coordinate with other agencies and international partners to counter maritime drug flows (achieving unity of effort)	1.2.1. Conduct joint operations (including shiprider operations) and exercises with other agencies and nations
		1.2.2. Assist partner nations in the development of capabilities to counter drug flows
		1.2.3. Develop MOUs and MOAs with other agencies and nations to enable more effective counterdrug operations
	1.3. Enable prosecution of smugglers	1.3.1. Collect and handle evidence for prosecution purposes
		1.3.2. Document case details and prepare case packages for prosecution
2. Increase intelligence and situational awareness of maritime drug flow	2.1. Conduct ISR to counter drug flows and trafficking networks	2.1.1. Investigate vessel and other physical evidence
		2.1.2. Interview apprehendees
		2.1.3. Deploy ISR systems
		2.1.4. Collect and integrate ISR systems' data
		2.2.1. Engage in interagency bodies and liaise with other agencies to enable information sharing and intelligence collaboration

Table D.2
Existing Metrics for Drug Interdiction

Outcome, Accomplishment, or Activity	Existing Metric	Source Document(s)	Comments on the Metric	Validity	Reliability	Feasibility	Comments on Ratings
1. Reduce the maritime flow of illegal drugs	Noncommercial maritime conveyance target	Coast Guard Office of Law Enforcement (CG-MLE)-1 FY 2015–2019 Law Enforcement Performance Targets		H	M	H	The quantity of drugs seized each year is easy to measure, but total drug flow must be estimated, limiting reliability
1.1. Deter, divert, detain, and disrupt maritime drug-smuggling flows	Coast Guard removal rates of cocaine as a percentage of CCDB removal rate targets	CG-MLE-1 FY 2015–2019 Law Enforcement Performance Targets	FY 2015–2019 targets are lower due to several factors	H	M	H	
	Coast Guard removal quantity targets in metric tons of cocaine, as determined by the CCDB working group	CG-MLE-1 FY 2015–2019 Law Enforcement Performance Targets; LANT-OPD FY 14		H	M	H	
	Coast Guard and partner agency removal rates of cocaine as a percentage in maritime transit zones	CG-MLE-1 FY 2015–2019 Law Enforcement Performance Targets	Percentage target represents Coast Guard and all blue force assets; Coast Guard's contributions not separated out	H	M	H	
1.1.5. Achieve visible or perceived presence	HC-130 hours to JIATF-S	LANT-OPD FY 14		H	H	H	
	Major cutter days to JIATF-S	LANT-OPD FY 14		H	H	H	
	Helicopter Interdiction Tactical Squadron days away from home station and days deployed onboard ship			H	H	H	

Table D.3
Potential Metrics for Drug Interdiction

Associated Activity, Accomplishment, or Outcome	Potential Metric	Comments on the Metric	Validity	Reliability	Feasibility	Comments on Ratings
1. Reduce the maritime flow of illegal drugs	Number of operations focused on consolidated priority organization targets, broken down by DTO		H	M	M	It may not be known which DTO a seizure affects
	Percentage of all maritime drug flow removed by Coast Guard, by drug type (marijuana, heroin, etc.)		H	M	M	
	Value of drugs ($) the Coast Guard removes from circulation each FY, by drug type		H	H	H	
	Quantity of drugs removed, in metric tons, by drug type		H	H	H	
1.1. Deter, divert, detain, and disrupt maritime drug-smuggling flows	Number of vessels observed changing course or reversing direction due to presence of Coast Guard assets		H	H	H	
	Number of events in which Coast Guard is unable to properly respond, detect, or monitor targets identified through all source intelligence reporting		H	H	H	
	Disruption rate for movement of controlled substances from noncommercial vessels in the maritime transit zone successfully detected by JIATF-S		H	H	H	
1.1.1. Detect and monitor	Number of drug vessels detected		H	H	H	
	Percentage of drug vessels detected, by vessel type, out of those transiting	Obviously an estimate because the denominator is not directly known	H	L	M	

Table D.3—Continued

Associated Activity, Accomplishment, or Outcome	Potential Metric	Comments on the Metric	Validity	Reliability	Feasibility	Comments on Ratings
	Percentage of detections due to maritime patrol aircraft	May not sum to 100% because these metrics are not mutually exclusive—more than one can contribute	M	H	H	
	Percentage of detections due to intelligence		M	M	H	
	Percentage of detections from land		M	H	H	
	Percentage of detections from vessels		M	H	H	
	Percentage of detections from non–Coast Guard assets		M	H	H	
1.1.2. Intercept	Number of drug vessels intercepted, by vessel type		H	H	H	
	Percentage of detected vessels that are intercepted, by vessel type		H	H	H	
1.1.3. Interdict	Number of drug vessels interdicted, by vessel type		H	H	H	
	Percentage of detected vessels that are interdicted, by vessel type		H	H	H	
1.1.4. Board and apprehend	Number of drug vessels boarded or apprehended, by vessel type		H	H	H	
	Percentage of drug vessels boarded out of those interdicted, by vessel type		H	H	H	
1.1.5. Achieve visible or perceived presence	HC-130 hours to JIATF-S		H	H	H	

Table D.3—Continued

Associated Activity, Accomplishment, or Outcome	Potential Metric	Comments on the Metric	Validity	Reliability	Feasibility	Comments on Ratings
	Major cutter days to JIATF-S		H	H	H	
	Helicopter Interdiction Tactical Squadron days away from home station and days deployed aboard ship		H	H	H	
	LEDET days in JIATF-S		H	H	H	
1.2. Support and coordinate with other agencies and international partners to counter maritime drug flows (achieving unity of effort)	Aggregate metric: average of metrics associated with subordinate activities		M	M	H	
	Number of mobile training teams sent to support theater security training initiatives each FY		M	H	H	
	Number of Coast Guard LEDETs assigned to USN combatants and ships from partner or allied nations		M	H	H	
1.2.1. Conduct joint operations (including shiprider operations) and exercises with other agencies and nations	Number of joint operations and exercises conducted with other agencies and nations per FY		M	H	H	
1.2.2. Assist partner nations in the development of capabilities to counter drug flows	Number of training events conducted in partner nations per FY		M	H	H	
1.2.3. Develop MOUs and MOAs with other agencies and nations to enable more effective counterdrug operations	Number of bilateral agreements on drug interdiction divided by desired number of such agreements (perhaps weighted based on relative importance of partner nation)		H	H	H	

Table D.3—Continued

Associated Activity, Accomplishment, or Outcome	Potential Metric	Comments on the Metric	Validity	Reliability	Feasibility	Comments on Ratings
	Percentage of cases in which bilateral agreements are invoked to enable operations		M	M	M	
	Percentage of cases in which actions under bilateral agreements substantially contribute to prosecution		M	M	M	
1.3. Enable prosecution of smugglers	Number of cases per year transferred to DoJ and other Law Enforcement partners for prosecution		H	H	H	
	Number of successful prosecutions of smugglers following Coast Guard apprehensions		H	H	M	Might require follow-up with DoJ because Coast Guard is not always involved in the prosecutions
1.3.1. Collect and handle evidence for prosecution purposes	Percentage of times in which evidence was unable to be used in case		H	H	M	"M" indicates that there is no formal system to retrieve this information from DoJ
1.3.2. Document case details and prepare case packages for prosecution	Percentage of time that all relevant and available types of evidence (e.g., physical evidence, telephone records, documents) were provided for prosecution		H	H	H	
2. Increase intelligence and situational awareness of maritime drug flow	Number of seizures of properly flagged (as opposed to stateless) vessels		M	H	H	
	Number of migrant vessels found with drugs		M	H	H	
	Weapons seizures with drug interdictions		M	H	H	
	Monetary seizures with drug interdictions		M	H	H	

Table D.3—Continued

Associated Activity, Accomplishment, or Outcome	Potential Metric	Comments on the Metric	Validity	Reliability	Feasibility	Comments on Ratings
2.1. Collect intelligence to combat drug networks	Number of events on which Coast Guard operational personnel have provided information to the Intelligence Coordination Center (ICC)		H	H	H	
	Number of incidents in which intelligence led Coast Guard assets to a drug vessel		H	H	H	
	Number of incidents in which actionable intelligence on a target was available, but the Coast Guard lacked either the assets with which to respond or the partner agency assets to act on the information		H	H	H	
	Asset hours dedicated to maritime smuggling monitoring using unmanned aircraft systems	Unmanned aircraft systems may be multimission	H	M	H	
	Percentage of times ISR information was successfully integrated into Coast Guard operational planning		M	M	M	
2.1.1. Investigate vessel and other physical evidence	Percentage of times evidence was gathered from seized drug vessels		H	H	H	
2.1.2. Interview apprehendees	Percentage of individuals interviewed out of those detained		H	H	H	
2.1.3. Deploy ISR systems	Number of incidents in which ISR was deployed		H	H	H	
2.1.4. Collect and integrate ISR systems' data	Number of times ISR information was used by operational personnel		M	M	M	

Table D.3—Continued

Associated Activity, Accomplishment, or Outcome	Potential Metric	Comments on the Metric	Validity	Reliability	Feasibility	Comments on Ratings
2.2. Share and collaboratively develop intelligence	Number of intelligence reports shared with Coast Guard field personnel on drug flow		H	H	H	
	Number of incidents in which intelligence on a target was available and the Coast Guard did not have assets available to send but was able to pass the information along to a partner agency that could act on the information		H	H	H	
2.2.1. Engage in interagency bodies and liaise with other agencies to enable information sharing and intelligence collaboration	Number of Coast Guard intelligence reports shared with interagency partners per FY		H	H	H	
	Number of full-time Coast Guard personnel addressing drug issues located at combatant commands, JIATF-S, interagency bodies, and as liaisons at other agencies		M	H	H	

APPENDIX E

Migrant Interdiction

Table E.1
Logic Model for Migrant Interdiction

Outcomes	Accomplishments	Activities
1. Deter undocumented maritime migration attempts	1.1. Raise public awareness of policies on, countermeasures against, and consequences of undocumented migration	1.1.1. Implement public-affairs campaigns in source countries in coordination with other agencies
		1.1.2. Implement public-affairs campaigns domestically in coordination with other agencies
		1.1.3. Publicize successful prosecutions and associated prison sentences for smugglers
	1.2. Demonstrate effective presence and capability to deter maritime migration	1.2.1. Ensure high visibility throughout high-traffic vectors
		1.2.2. Ensure periodic visibility in low-traffic vectors
		1.2.3. Conduct high-profile mass-migration exercises
	1.3. Enable prosecution of smugglers	1.3.1. Identify, apprehend, and transfer smugglers
		1.3.2. Prepare case packages for prosecution of smugglers
2. Prevent undocumented migrants from reaching U.S. territory via maritime routes	2.1. Build interagency and international partnerships to share information, coordinate plans, and operate cohesively	2.1.1. Engage in interagency bodies, leverage interagency centers, exercise with other agencies, and operate alongside other agencies
		2.1.2. Engage bilaterally and multilaterally to enable information sharing, shipriding, training, enhancement of partner-nation capabilities, and cooperative efforts
		2.1.3. Continue development of bilateral agreements to facilitate the swift repatriation of migrants after interdiction
	2.2. Achieve enhanced situational awareness of and knowledge on undocumented migration	2.2.1. Enhance abilities to predict emerging migration threats and new smuggling routes
		2.2.2. Estimate migrant flow
	2.3. Detect, interdict, and repatriate migrants (or enable repatriation)	2.3.1. Detect migrants
		2.3.2. Interdict migrants
		2.3.3. Repatriate migrants
		2.3.4. Rescue migrants from overloaded and/or unseaworthy vessels
		2.3.5. Provide humanitarian aid to interdicted migrants
		2.3.6. Provide access to protection screening in accordance with law, policy, and agreements

Table E.2
Existing Metrics for Migrant Interdiction

Outcome, Accomplishment, or Activity	Existing Metric	Source Document(s)	Comments on the Metric	Validity	Reliability	Feasibility	Comments on Ratings
1. Deter undocumented maritime migration attempts	Number of undocumented migrants attempting to enter the U.S. via maritime routes	Michel, 2014, p. 15; 2014 Third Quarter Performance Review, p. 26	Data source: Maritime Intelligence Fusion Center (MIFC) (sorted by nationality)	H	M	M	Reliability and feasibility are medium because of uncertainty about the number of unseen migrants trying to enter the United States; it would be difficult to accurately determine this information
1.2. Demonstrate effective presence and capability to deter maritime migration	Asset-hours utilized for migrant interdiction	2014 Third Quarter Performance Review, p. 10	Resources: aircraft, cutter, boat	M	H	H	Presence and activity are not distinguished
1.3.1. Identify, apprehend, and transfer smugglers	Number of smugglers detained	USCG, 2014b	Not explicitly stated as a metric	H	H	H	
2. Prevent undocumented migrants from reaching U.S. territory via maritime routes	Total known maritime undocumented migrant flow, with breakdown by interdicting or apprehending agency	2014 Third Quarter Performance Review, p. 25	Compares FY13 and 14: Interdicted by Coast Guard, shoreside apprehensions, interdicted by partner nations	H	M	M	Reliability and feasibility are medium because other agencies or nations might not share information, might not measure items the same way, or might not share the same quality standards
2.3.2. Interdict migrants	Number of undocumented migrants who attempt to enter the United States via maritime routes and are interdicted	Michel, 2014, p. 15—this is also listed on p. 4 as a management measure; 2014 Third Quarter Performance Review, p. 8	Data source: MIFC (reported by area); compares across years and lists a target to achieve	H	M	H	Reliability is medium because Coast Guard documents state that there is a data discrepancy between the statistics in MISLE and those in ICC's database

Table D.2—Continued

Outcome, Accomplishment, or Activity	Existing Metric	Source Document(s)	Comments on the Metric	Validity	Reliability	Feasibility	Comments on Ratings
	Number of undocumented migrants who attempt to enter the United States via maritime routes and are interdicted by the Coast Guard, divided by area	LANTAREA OPD, April 2014, p. 7	Would appear to be a subset of information gathered for overall Coast Guard statistics	H	H	H	
	Percentage of undocumented migrants who attempt to enter the United States via maritime routes and are interdicted	Michel, 2014, p. 15; 2014 Third Quarter Performance Review, p. 26	Data source: MIFC (reported by area)	H	M	M	Reliability and feasibility are medium because the denominator is uncertain and difficult to attain (see above) and because Cost Guard documents state that there is a data discrepancy between the statistics in MISLE and those in ICC's database
	Percentage of undocumented migrants attempting to enter the United States via maritime routes and are interdicted by the Coast Guard	Michel, 2014 p. 15; 2014 Third Quarter Performance Review, p. 26	Data source: MIFC, but not automatically calculated (reported by area)	H	M	M	Reliability and feasibility are low because the denominator is uncertain and difficult to attain (see above) and because two separate documents cite the same database but FY13 Operational Performance states the measure is not automatically calculated
N/A	Number of alien migrant interdiction operations cases requiring elevated use of force	Neffenger, 2013b, p. 11	Recently dropped; previously MISLE (reported by area)	N/A	H	H	

Table E.3
Potential Metrics for Migrant Interdiction

Associated Activity, Accomplishment, or Outcome	Potential Metric	Comments on the Metric	Validity	Reliability	Feasibility	Comments on Ratings
1. Deter undocumented maritime migration attempts	Estimated cost of smuggling per migrant, by country		M	M	M	Can be measured by asking migrants, although they may not share the truth (or any value at all); more-accurate numbers may depend on intelligence
1.1. Raise public awareness of policies on, countermeasures against, and consequences of undocumented migration	Aggregate metric: average of metrics associated with subordinate activities		M	M	H	
1.1.1. Implement public-affairs campaigns in source countries in coordination with other agencies	Number of public-affairs campaigns being conducted in source countries		M	H	H	
	Forms of communication that public-affairs campaigns in source countries are using (e.g., radio, television, Internet)		L	H	H	
	Estimated percentage of relevant audience being reached by public-affairs campaigns in source countries	Accuracy requires surveys	H	L	L	
	Degree to which public-affairs campaigns in source countries influence migration decisions	Accuracy requires surveys	H	L	L	
1.1.2. Implement public-affairs campaigns domestically in coordination with other agencies	Number of public-affairs campaigns being conducted by the United States		M	H	H	
	Forms of communication that public-affairs campaigns in the United States are using (e.g., radio, television, Internet)		L	H	H	

Table D.3—Continued

Associated Activity, Accomplishment, or Outcome	Potential Metric	Comments on the Metric	Validity	Reliability	Feasibility	Comments on Ratings
	Estimated percentage of relevant U.S. audience being reached by public-affairs campaigns in source countries	Accuracy requires surveys	H	L	L	
	Degree to which public-affairs campaigns within U.S. influence migration decisions abroad	Accuracy requires surveys	H	L	L	
1.1.3. Publicize successful prosecutions and associated prison sentences for smugglers	Number of prosecutions of smugglers publicized on U.S. government websites and/or as part of public-affairs campaigns		H	H	H	
1.2. Demonstrate effective presence and capability to deter maritime migration	Asset-hours, by asset type, used for this mission	Discerning which hours were spent on migrant interdiction, as opposed to drug interdiction or other missions, is challenging; platforms are often performing multiple missions at a time or shifting rapidly from one mission to another	M	M	M	
1.2.1. Ensure high visibility throughout high-traffic vectors	Number of asset-hours, by type, dedicated to high-traffic vectors and the times of day or night engaged in this mission		M	M	M	
1.2.2. Ensure periodic visibility in low-traffic vectors	Number of asset-hours, by type, dedicated to low-traffic vectors and the times of day or night engaged in this mission		M	M	M	
1.2.3. Conduct high-profile mass-migration exercises	Number of mass-migration exercises conducted each FY		H	H	H	
1.3. Enable prosecution of smugglers	Number of smugglers prosecuted		H	H	H	

Table D.3—Continued

Associated Activity, Accomplishment, or Outcome	Potential Metric	Comments on the Metric	Validity	Reliability	Feasibility	Comments on Ratings
	Percentage of apprehended smugglers turned over to other agencies and prosecuted		H	H	H	
1.3.1. Identify, apprehend, and transfer smugglers	Percentage of interdictions that appear to involve smugglers in which the smugglers are identified		H	H	H	
1.3.2. Prepare case packages for prosecution of smugglers	Percentage of smugglers apprehended for whom a completed case package was prepared		H	H	H	
2. Prevent undocumented migrants from reaching U.S. territory via maritime routes	Percentage of undocumented migrants who attempt to enter the United States via maritime routes and are interdicted, by nationality		H	H	H	
	Percentage of undocumented migrants attempting to enter the United States via maritime routes that are interdicted by the Coast Guard, by nationality		H	M	M	
	Number of undocumented migrants who attempt to enter the United States via maritime routes and are interdicted by the Coast Guard, by nationality, district, and area		H	H	H	
2.1. Build interagency and international partnerships to share information, coordinate plans, and operate cohesively	Number of high-level Coast Guard meetings with interagency and international partners about migrants		M	H	H	
	Number of interagency agreements (e.g., MOA, MOU) on migrants, divided by the number desired		M	H	H	

Table D.3—Continued

Associated Activity, Accomplishment, or Outcome	Potential Metric	Comments on the Metric	Validity	Reliability	Feasibility	Comments on Ratings
	Number of bilateral or multilateral agreements, divided by the number desired		M	H	H	
2.1.1. Engage in interagency bodies, leverage interagency centers, exercise with other agencies, and operate alongside other agencies	Percentage of interagency bodies and centers dealing with migrants in which the Coast Guard has appropriate representation		M	H	H	
	Number of full-time Coast Guard liaison personnel addressing migrants at other agencies or interagency centers or bodies, divided by desirable number of liaison personnel (or divided by the number of agencies plus the number of bodies or centers)		M	H	H	
	Number of other agencies' and interagency liaison personnel addressing migrants at relevant Coast Guard commands		M	H	H	
	Number of Coast Guard shipriders aboard other agencies' vessels conducting at least some migrant interdiction		M	H	H	
	Number of other agencies' shipriders aboard Coast Guard vessels conducting at least some migrant interdiction		M	H	H	
	Number of multiagency migrant exercises in which the Coast Guard participates		M	H	H	

Table D.3—Continued

Associated Activity, Accomplishment, or Outcome	Potential Metric	Comments on the Metric	Validity	Reliability	Feasibility	Comments on Ratings
	Percentage of multiagency migrant exercises in which the Coast Guard participates		M	H	H	
	Number of migrant operations conducted in cooperation with other agencies		M	H	H	
2.1.2. Engage bilaterally and multilaterally to enable information sharing, shipriding, training, enhancement of partner-nation capabilities, and cooperative efforts	Number of full-time Coast Guard liaison personnel working with other countries regarding migrants		M	H	H	
	Number of other nations' liaison personnel at relevant Coast Guard commands dealing with migrants		M	H	H	
	Number of Coast Guard shipriders aboard other nations' vessels dealing with migrants	Could be characterized in terms of either shiprider-days over a period of time or average numbers over time	M	H	H	
	Number of other nations' shipriders aboard Coast Guard vessels dealing with migrants	Could be characterized in terms of either shiprider-days over a period of time or average numbers over time	M	H	H	
	Number of bilateral or multilateral training events or exercises regarding migrants to which the Coast Guard contributes		M	H	H	
	Percentage of annual bilateral or multilateral training events or exercises regarding migrants to which the Coast Guard contributes		M	H	H	

Table D.3—Continued

Associated Activity, Accomplishment, or Outcome	Potential Metric	Comments on the Metric	Validity	Reliability	Feasibility	Comments on Ratings
	Number of migrant operations conducted in cooperation with other nations		M	H	H	
2.1.3. Continue development of bilateral agreements to facilitate the swift repatriation of migrants after interdiction	Number of bilateral agreements on repatriation divided by the desired number of such agreements	Could be weighted according to the relative importance of the partner nation	H	H	H	
2.2. Achieve enhanced situational awareness and knowledge regarding undocumented migration	Number of intelligence reports provided to the field regarding undocumented migration		M	H	H	
2.2.1. Enhance abilities to predict emerging migration threats and new smuggling routes	Number of full-time Coast Guard personnel devoted to analysis of emerging migration threats and new routes		L	H	H	
	Number of Coast Guard analytical reports concerning migration threats and routes		M	H	H	
	Ratings by other intelligence professionals of the quality of Coast Guard analysis of emerging migration threats and routes		M	M	M	
	Ratings by Coast Guard operators of the quality of Coast Guard analysis of emerging migration threats and routes		M	M	M	
2.2.2. Estimate migrant flow	Yearly report on the estimated migrant flow for the following year		M	H	H	

Table D.3—Continued

Associated Activity, Accomplishment, or Outcome	Potential Metric	Comments on the Metric	Validity	Reliability	Feasibility	Comments on Ratings
	Percentage of periodic reporting requirements (e.g., annually, quarterly) detailing migrant flows met		L	H	H	
	Retrospective accuracy of estimates of migrant flow, both overall and by nationality		H	M	M	
2.3. Detect, interdict, and repatriate migrants (or enable repatriation)	Number of incidents in which intelligence cues regarding a target were available and the Coast Guard was able to send its own assets, pass the information on to a partner agency for action, or had no assets or partner agency available to act on the information		H	H	H	
	Percentage of interdicted individuals who are recidivists	Requires biometrics	M	M	M	
	Percentage of interdicted individuals with nonimmigration criminal records		M	M	M	
2.3.1. Detect migrants	Number of vessels detected by the Coast Guard		H	H	H	
	Number of vessels detected by the Coast Guard, divided by the number of vessels estimated to have transited (based on other migrant-flow data)		H	L	M	

Table D.3—Continued

Associated Activity, Accomplishment, or Outcome	Potential Metric	Comments on the Metric	Validity	Reliability	Feasibility	Comments on Ratings
	Number of vessels detected by all agencies, including the Coast Guard, divided by the number of vessels estimated to have transited (based on other migrant-flow data)		H	L	M	
2.3.2. Interdict migrants	Number of undocumented migrants who attempt to enter the United States via maritime routes and are interdicted by the Coast Guard, by nationality	Could also be parsed by vector of transit	H	H	H	
	Number of undocumented migrants who attempt to enter the United States via maritime routes and are interdicted by all agencies, by nationality	Could also be parsed by vector of transit	H	H	H	
	Estimated percentage of undocumented migrants who attempt to enter the United States via maritime routes and are interdicted by the Coast Guard, by nationality	Could also be parsed by vector of transit	H	L	M	
	Estimated percentage of undocumented migrants who attempt to enter the United States via maritime routes and are interdicted by all agencies, by nationality	Could also be parsed by vector of transit	H	L	M	
2.3.3. Repatriate migrants	Number of migrants the Coast Guard was able to repatriate within a given time frame (hours or days), by nationality		H	H	H	

Table D.3—Continued

Associated Activity, Accomplishment, or Outcome	Potential Metric	Comments on the Metric	Validity	Reliability	Feasibility	Comments on Ratings
	Percentage of migrants (less those claiming credible fear of return whose applications are deemed eligible for further review) that the Coast Guard is able to swiftly repatriate, by nationality		H	H	H	
	Percentage of migrants claiming credible fear of return that are deemed ineligible, but are not repatriated within a given period (e.g., one day, week, or month), by nationality		H	H	H	
	Number of migrants claiming credible fear of return, by nationality		L	H	H	
	Percentage of migrants claiming credible fear of return, by nationality		L	H	H	
2.3.4. Rescue migrants from overloaded and/or unseaworthy vessels	Number of vessels rescued or encountered by the Coast Guard, by nationality		H	H	H	
	Number of individuals in need of rescue from vessels rescued by the Coast Guard		H	H	H	
	Percentage of estimated number of people in need of rescue from vessels rescued by the Coast Guard, by nationality		H	H	H	
	Percentage of estimated number of people in need of rescue from vessels rescued by all agencies, including the Coast Guard		H	H	H	

Table D.3—Continued

Associated Activity, Accomplishment, or Outcome	Potential Metric	Comments on the Metric	Validity	Reliability	Feasibility	Comments on Ratings
2.3.5. Provide humanitarian aid to interdicted migrants	Percentage of migrants who needed food, water, and/or other humanitarian supplies		H	H	H	
	Number of documented instances of mistreatment divided by number of migrants interdicted	This number should be zero and will almost universally be zero; however, the Coast Guard should report this number as a good-news story, particularly given current concerns about U.S. treatment of individuals in custody	H	H	H	
2.3.6. Provide access to protection screening in accordance with law, policy, and agreements	Percentage of migrants who received protection screening out of the total number of migrants, by nationality		H	H	H	

APPENDIX F

Defense Readiness

Table F.1
Logic Model for Defense Readiness

Outcomes	Accomplishments	Activities
1. Fulfill DoD requirements	1.1. Provide teams and support for counterterrorism; counterdrug; and coastal, port, and harbor security operations	1.1.1. Provide VBSS team support
		1.1.2. Provide RAID support
		1.1.3. Provide deployable force packages (e.g., K9, RWAI, MSRT, MSST)
		1.1.4. Provide AUF capability support
		1.1.5. Provide PSU support
	1.2. Conduct operations for combatant commanders	1.2.1. Conduct TSC
		1.2.2. Conduct coastal air and surface operations
		1.2.3. Conduct port and harbor operations
		1.2.4. Conduct Arctic and Antarctic coastal sea control operations
2. Maintain ability to respond to calls for forces	2.1. Maintain readiness of forces	2.1.1. Conduct independent and joint military training
		2.1.2. Conduct independent and joint military exercises
	2.2. Maintain interoperability with DoD forces	2.2.1. Conduct joint military operations
		2.2.2. Conduct joint military communications
3. Fulfill homeland defense requirements	3.1. Conduct interception and interdiction operations	3.1.1. Execute RWAI duties
		3.1.2. Provide cutters and boats for interception and/or interdiction operations
		3.1.3. Provide LEDET support
		3.1.4. Provide MOTR support
	3.2. Conduct military environmental response operations	3.2.1. Conduct spill and release notifications
		3.2.2. Conduct spill and release incident management
		3.2.3. Conduct spill and release containment and cleanup operations
		3.2.4. Conduct pollution incident investigation and documentation

Table F.2
Existing Metrics for Defense Readiness

Outcome, Accomplishment, or Activity	Existing Metric	Source Document(s)	Comments on the Metric	Validity	Reliability	Feasibility	Comments on Ratings
1. Fulfill DoD requirements 3. Fulfill homeland defense requirements 3.1.3. Provide LEDET support	Percentage of time patrol boats are available to fulfill DoD requirements	Michel, 2014	Data source: CG-RACE Target: 100%	L	H	H	Metrics are asset-centric, while outcomes, accomplishments, and activities are mission-centric, which is why most metrics are mapped to multiple items, limiting their validity in measuring any one of them
	Percentage of time major cutters are available to fulfill DoD requirements	Michel, 2014	Data source: CG-RACE Target: 100%	L	H	H	
1. Fulfill DoD requirements 1.2. Conduct operations for combatant commanders 1.2.1. Conduct TSC 3. Fulfill homeland defense requirements	Percentage of time deployed PSUs are available to fulfill DoD requirements	Michel, 2014	Data source: CG-RACE Target: 100%	L	H	H	
	Percentage of time ready-to-deploy PSUs are available to fulfill DoD requirements	Michel, 2014	Data source: CG-RACE Target: 100%	L	H	H	
2. Maintain ability to respond to calls for forces 2.1. Maintain readiness of forces 2.1.2. Conduct independent and joint military exercises 2.2.1. Conduct joint military operations	Major cutter exercise support for theater security cooperation	Michel, 2014	Area cutter schedules Target: 85 days	L	H	H	

Table F.2—Continued

Outcome, Accomplishment, or Activity	Existing Metric	Source Document(s)	Comments on the Metric	Validity	Reliability	Feasibility	Comments on Ratings
1. Fulfill DoD requirements 1.1. Provide teams and support for counterterrorism; counterdrug; and coastal, port, and harbor security operations 1.2. Conduct operations for combatant commanders 1.2.2. Conduct coastal air and surface operations 1.2.3. Conduct port and harbor operations 3. Fulfill homeland defense requirements 3.1. Conduct interception and interdiction operations 3.1.2. Provide cutters and boats for interception and/or interdiction operations	WPB overseas contingency operations (OCO) support—U.S. Central Command	Michel, 2014	Data source: CGBI Target: 6.0 coverage	L	H	H	

Table F.2—Continued

Outcome, Accomplishment, or Activity	Existing Metric	Source Document(s)	Comments on the Metric	Validity	Reliability	Feasibility	Comments on Ratings
1. Fulfill DoD requirements 1.1. Provide teams and support for counterterrorism; counterdrug; and coastal, port, and harbor security operations 1.1.5. Provide PSU support 1.2. Conduct operations for combatant commanders 2.3. Conduct port and harbor operations 3. Fulfill homeland defense requirements	PSU OCO support—USSOUTHCOM	Michel, 2014	Data source: Coast Guard Office of Counterterrorism and Defense Operations Policy (CG-ODO) Target: 1.0 coverage	L	H	H	
1. Fulfill DoD requirements 1.1. Provide VBSS team support 2. Maintain ability to respond to calls for forces 2.2.1. Conduct joint military operations 3. Fulfill homeland defense requirements	VBSS OCO support—U.S. Central Command	Michel, 2014	Data source: CG-ODO Target: 0.5 coverage	L	H	H	

Table F.2—Continued

Outcome, Accomplishment, or Activity	Existing Metric	Source Document(s)	Comments on the Metric	Validity	Reliability	Feasibility	Comments on Ratings
1. Fulfill DoD requirements 1.1. Provide teams and support for counterterrorism; counterdrug; and coastal, port, and harbor security operations 1.1.1. Provide VBSS team support 1.1.3. Provide deployable force packages (e.g., K9, RWAI, MSRT, MSST) 3. Fulfill homeland defense requirements 3.1. Conduct interception and interdiction operations 3.1.3. Provide LEDET support	LEDET support—USSOUTHCOM	Michel, 2014	Data source: CG-ODO Target: 5.0 coverage	L	H	H	

Table F.2—Continued

Outcome, Accomplishment, or Activity	Existing Metric	Source Document(s)	Comments on the Metric	Validity	Reliability	Feasibility	Comments on Ratings
1. Fulfill DoD requirements 1.1. Provide teams and support for counterterrorism; counterdrug; and coastal, port, and harbor security operations 1.1.2. Provide RAID support 1.1.3. Provide deployable force packages (e.g., K9, RWAI, MSRT, MSST) 1.2.3. Conduct port and harbor operations 2. Maintain ability to respond to calls for forces 2.1. Maintain readiness of forces 2.2. Maintain interoperability with DoD forces 2.2.1. Conduct joint military operations	RAID team support—U.S. Transportation Command	Michel, 2014	Data source: CG-ODO Target: 1.0 coverage	L	H	H	
1. Fulfill DoD requirements 1.2. Conduct operations for combatant commanders 1.2.4. Conduct Arctic and Antarctic coastal sea control operations	420-foot Healy-class Icebreaker support—U.S. Pacific Command	Michel, 2014	Data source: CG-ODO Target: 0.5 coverage	L	H	H	

Table F.2—Continued

Outcome, Accomplishment, or Activity	Existing Metric	Source Document(s)	Comments on the Metric	Validity	Reliability	Feasibility	Comments on Ratings
1. Fulfill DoD requirements 2. Maintain ability to respond to calls for forces 3. Fulfill homeland defense requirements	Out of hemisphere deployment days completed	CGBI out of hemisphere	CGBI measure ID: 21476	L	H	H	
1. Fulfill DoD requirements 1.1. Provide teams and support for counterterrorism; counterdrug; and coastal, port, and harbor security operations 1.1.1. Provide VBSS team support 1.2. Conduct operations for combatant commanders 1.2.1. Conduct TSC 3. Fulfill homeland defense requirements	PSU, VBSS, and WPB OCO support	CGBI	CGBI measure ID: 22135, 22134, 21477	L	H	H	
2. Maintain ability to respond to calls for forces 2.1. Maintain readiness of forces 2.1.1. Conduct independent and joint military training 2.2. Maintain interoperability with DoD forces	PSU ready for operation results;	CGBI	CGBI measure ID: 22136, 22138, 22137	M	M	M	

Table F.2—Continued

Outcome, Accomplishment, or Activity	Existing Metric	Source Document(s)	Comments on the Metric	Validity	Reliability	Feasibility	Comments on Ratings
2. Maintain ability to respond to calls for forces 2.1. Maintain readiness of forces 2.1.1. Conduct independent and joint military training 2.2. Maintain interoperability with DoD forces	WPB special emergency operations and procedures results	CGBI	CGBI measure ID: 22136, 22138, 22137	M	M	M	
2. Maintain ability to respond to calls for forces 2.1. Maintain readiness of forces 2.1.1. Conduct independent and joint military training 2.2. Maintain interoperability with DoD forces	High-Endurance Cutter/18-foot National Security Cutter tailored ship's training availability results	CGBI	CGBI measure ID: 22136, 22138, 22137	M	M	M	
2.1. Maintain readiness of forces	Readiness of deployable units for defense operations	CGBI	CGBI measure ID: 21294	M	L	L	
1. Fulfill DoD requirements 2. Maintain ability to respond to calls for forces 3. Fulfill homeland defense requirements	Percentage of time High-Endurance Cutter achieves command and control (C2) readiness	Albero, 2014	Target: 33%	L	L	L	
	Percentage of time WPB achieves C2 readiness	Albero, 2014	Target: 26.3%	L	L	L	
	Percentage of time PSU achieves C2 readiness	Albero, 2014	Target: 50%	L	L	L	

Table F.2—Continued

Outcome, Accomplishment, or Activity	Existing Metric	Source Document(s)	Comments on the Metric	Validity	Reliability	Feasibility	Comments on Ratings
1.1.3. Provide deployable force packages (e.g., K9, RWAI, MSRT, MSST) 3.1.1. Execute RWAI duties	Deployed RWAI days	Interview		M	H	H	

Table F.3
Potential Metrics for Defense Readiness

Associated Activity, Accomplishment, or Outcome	Potential Metric	Comments on the Metric	Validity	Reliability	Feasibility	Comments on Ratings
1. Fulfill DoD requirements	Weighted average of metrics for subordinate accomplishments	Weights within average reflect relative importance	H	H	H	
1.1. Provide teams and support for counterterrorism, counterdrug, and coastal, port, and harbor security operations	Weighted average of metrics for subordinate activities	Weights within average reflect relative importance	H	H	H	
1.1.1. Provide VBSS team support	Percentage of requests met		H	H	H	
1.1.2. Provide RAID support	Percentage of requests met		H	H	H	
1.1.3. Provide deployable force packages (e.g., K9, RWAI, MSRT, MSST)	Percentage of requests met		H	H	H	
1.1.4. Provide AUF capability support	Percentage of requests met		H	H	H	
1.1.5. Provide PSU support	Percentage of requests met		H	H	H	
1.2. Conduct operations for combatant commanders	Weighted average of metrics for subordinate activities	Weights within average reflect relative importance	H	H	H	
1.2.1. Conduct TSC	Percentage of requests met		H	H	H	
1.2.2. Conduct coastal air and surface operations	Percentage of requests met		H	H	H	
1.2.3. Conduct port and harbor operations	Percentage of requests met		H	H	H	
1.2.4. Conduct Arctic and Antarctic coastal sea control operations	Percentage of requests met		H	H	H	

Table F.3—Continued

Associated Activity, Accomplishment, or Outcome	Potential Metric	Comments on the Metric	Validity	Reliability	Feasibility	Comments on Ratings
2. Maintain ability to respond to calls for forces	DoD ratings on the Coast Guard's ability to respond, on a 0–10 scale	This could be achieved by surveying relevant DoD commanders	M	M	L	
	Coast Guard ratings regarding the Coast Guard's ability to respond, on a 0–10 scale	This could be achieved by surveying relevant Coast Guard commanders	M	M	L	
2.1. Maintain readiness of forces	Number of military training events and exercises in which the Coast Guard participates		M	H	H	
	Percentage of desirable military training events and exercises in which the Coast Guard participates		M	M	H	
2.1.1. Conduct independent and joint military training	Number of military training events in which the Coast Guard participates		M	H	H	
	Percentage of desirable military training events in which the Coast Guard participates		M	M	H	
2.1.2. Conduct independent and joint military exercises	Number of military exercises in which the Coast Guard participates		M	H	H	
	Percentage of desirable military exercises in which the Coast Guard participates		M	M	H	
2.2. Maintain interoperability with DoD forces	DoD ratings regarding Coast Guard interoperability, on a 0–10 scale	Could be achieved by surveying relevant DoD commanders	M	M	L	
	Coast Guard ratings regarding interoperability, on a 0–10 scale	Could be achieved by surveying relevant Coast Guard commanders	M	M	L	

Table F.3—Continued

Associated Activity, Accomplishment, or Outcome	Potential Metric	Comments on the Metric	Validity	Reliability	Feasibility	Comments on Ratings
2.2.1. Conduct joint military operations	Number of joint military operations conducted		M	H	H	
	Percentage of requests met		M	H	H	
2.2.2. Conduct joint military communications	Number of instances of the Coast Guard conducting joint communications with elements of DoD		M	M	H	
	Percentage of relevant DoD communication systems that the Coast Guard also uses		M	M	H	
3. Fulfill homeland defense requirements	Weighted average of percentage of requests met for interception, interdiction, and military environmental response operations		M	H	H	
3.1. Conduct interception and interdiction operations	Weighted average of metrics for subordinate activities	Weights within average reflect relative importance	H	H	H	
3.1.1. Execute RWAI duties	Percentage of requests met		H	H	H	
3.1.2. Provide cutters and boats for interception and/or interdiction operations	Percentage of requests met		H	H	H	
3.1.3. Provide LEDET support	Percentage of requests met		H	H	H	
3.1.4. Provide MOTR support	Percentage of requests met		H	H	H	
3.2. Conduct military environmental response operations	Number of military environmental response operations to which the Coast Guard contributes		M	H	H	
	Percentage of military environmental response operations to which the Coast Guard contributes		M	H	H	

Table F.3—Continued

Associated Activity, Accomplishment, or Outcome	Potential Metric	Comments on the Metric	Validity	Reliability	Feasibility	Comments on Ratings
	Percentage of requests met		M	H	H	
3.2.1. Conduct spill and release notifications	Percentage of appropriate agencies notified within designated timespan		M	M	M	
	Number of communications channels used (radio, television, Internet, etc.) to communicate with public as needed		L	H	H	
	Estimated percentage of relevant public reached	This requires surveys after the fact and estimates of how many were affected.	M	L	L	
3.2.2. Conduct spill and release incident management	Percentage of material recovered	This will be influenced by many factors beyond the Coast Guard's control	L	M	M	
	Number of safety incidents arising during course of response	This will be influenced by many factors beyond the Coast Guard's control	L	M	M	

APPENDIX G

Ice Operations

Table G.1
Logic Model for Ice Operations

Outcomes	Accomplishments	Activities
1. Facilitate safe and efficient activity in icy waters while reducing the risk and severity of ice-related incidents	1.1. Enable maritime commerce to proceed safely without significant delay despite icy conditions	1.1.1. Monitor and report ice conditions on coastal and inland waterways
		1.1.2. Ensure that vessels operating independently in ice have sufficient horsepower and hull reinforcement
		1.1.3. Break ice to allow certain commercial vessels to utilize iced waterways, as necessary
		1.1.4. Coordinate with other public and private icebreaking entities to ensure adherence and appropriate modifications to standards
	1.2. Enable access to icebound polar regions	1.2.1. Deploy polar-class icebreakers to polar regions
	1.3. Minimize human casualties, property damage, and environmental damage from vessels beset by ice	1.3.1. Rescue vessels beset by ice
	1.4. Minimize human casualties and property damage from flooding due to ice jams	1.4.1. Break ice jams that pose a flooding hazard

Table G.2
Existing Metrics for Ice Operations

Outcome, Accomplishment, or Activity	Existing Metric	Source Document(s)	Comments on the Metric	Validity	Reliability	Feasibility	Comments on Ratings
1.1.3. Break ice to allow certain commercial vessels to utilize iced waterways as necessary	Number of days high-priority waterways are closed to commerce because of ice	Neffenger, 2013b, Enclosure 2, p. 13; Servidio, 2013a, pp. 50–54; Albero, 2014, p. 11; OPAR, p. 7	No longer tracked	H	H	H	
	Domestic icebreaking waterway availability	Albero, 2014, p. 11; OPAR, p. 7	Other ice operations measures in OPAR	H	H	H	
	Percentage of time high-priority waterways in the Great Lakes and along the eastern seaboard are open during the ice season	Michel, 2014, Enclosure 3; Presidential Policy Directive (PPD), pp. 49–57	Replaced the "number of days" metric from FY14 and before Comments from Michel, 2014: This is calculated by dividing the number of hours of Tier One waterways available by the total hours available during the ice season. This is reported by units with Domestic Icebreaking (DOMICE) assets in D1, D5, and D9.	H	H	H	
	Requests for domestic icebreaking met	Albero, 2014, p. 11; OPAR, p. 7; Servidio, 2013a, pp. 50–54; Michel, 2014, Enclosure 3 PPD; pp. 49–57	Other ice operations measures in OPAR Comments from Michel, 2014: This is calculated by dividing the number of requests met by the total number of requests made during the ice season. This is reported by units with DOMICE assets in D1, D5, and D9.	H	H	H	

Table G.2—Continued

Outcome, Accomplishment, or Activity	Existing Metric	Source Document(s)	Comments on the Metric	Validity	Reliability	Feasibility	Comments on Ratings
	Domestic icebreaking cutter capacity	Michel, 2014, Enclosure 3 PPD, pp. 49–57	Other ice operations measures in OPAR Comments from Michel, 2014: This captures a DOMICE cutter's capacity rate over time (the time it is not in unscheduled Charlie status compared to total time to perform icebreaking or other missions during the ice season from 15 Dec to 20 Apr. This is reported by units with DOMICE assets in D1, D5, and D9.	H	H	H	

Table G.3
Potential Metrics for Ice Operations

Associated Activity, Accomplishment, or Outcome	Potential Metric	Comments on the Metric	Validity	Reliability	Feasibility	Comments on Ratings
1. Facilitate safe and efficient activity in icy waters while reducing the risk and severity of ice-related incidents	Number of CAGSM incidents in which ice is a factor, normalized by the severity of the ice season (freezing degree days, percentage ice coverage in navigable waterways, or ice thickness in navigable waterways) and per transit, vessel-mile, ton-mile, or dollar-mile	This includes only incidents caused or exacerbated by icy conditions. These include CAGSMs, such as vessels beset by ice	H	H	M	
	The economic impact of commerce facilitated normalized by the ice severity	New metric suggested during a round of feedback.	H	M	M	Economic impact measures are always subject to debate. The core of the debate typically centers on what should be included in the impact and what should be excluded. Also, how different is impact mitigated from impact delayed? There is a feasibility trade-off that should also be considered. The more precisely the data reflect reality, the more likely the data are proprietary or private (e.g., cargo values, deal terms, salaries). Therefore, the feasibility of obtaining accurate economic impact data is "M." Also, the severity of the ice season will vary from one mile of waterway to the next and would be prohibitive to measure accurately. Therefore, the reliability of this proposed measure is "M."

Table G.3—Continued

Associated Activity, Accomplishment, or Outcome	Potential Metric	Comments on the Metric	Validity	Reliability	Feasibility	Comments on Ratings
1.1. Enable maritime commerce to proceed safely without significant delay despite icy conditions	Number of transits, vessel-miles, ton-miles, or dollar-miles, normalized by severity of the ice season (freezing-degree days, percentage ice coverage in navigable waterways, or ice thickness in navigable waterways)		M	M	M	
	Timeliness of reporting ice conditions to mariners (average delay from time of observation to time of dissemination)		H	H	M	
1.1.1. Monitor and report ice conditions on coastal and inland waterways	Accuracy of reported ice conditions, determined through audits or by comparison with subsequent reports		H	M	M	
	Average delay in public dissemination after change in conditions observed or reported		M	M	M	
	Accessibility of reported ice conditions as compared to best practices	Done correctly, this measure would encourage the Coast Guard to be responsive to the information pathways mariners use, e.g., smartphone apps	M	H	M	
1.1.2. Ensure that vessels operating independently in ice have sufficient horsepower and hull reinforcement	Number of vessels beset by ice normalized by the severity of the conditions	As above, severity could include freezing degree days, percentage of coverage in navigable waterways, and ice thickness	M	H	M	Validity and feasibility rate as "M" because the severity of the conditions at any one time or place, or overall, would be difficult to determine

Table G.3—Continued

Associated Activity, Accomplishment, or Outcome	Potential Metric	Comments on the Metric	Validity	Reliability	Feasibility	Comments on Ratings
	Duration and severity of the horsepower and hull thickness restrictions normalized by the duration and severity of restrictions the maritime community feels is appropriate	The maritime community's opinion about the appropriateness of the restrictions could be collected in group forums or via surveys after the ice season is finished	M	M	L	Rates "L" for feasibility because the denominator would be difficult to collect accurately
1.1.3. Break ice to allow certain commercial vessels to utilize iced waterways as necessary	None needed—covered by existing metric					
1.1.4. Coordinate with other public and private icebreaking entities to ensure adherence and appropriate modifications to standards	Percentage of non–Coast Guard icebreaking vessels that meet standards		H	H	H	
	Number of CAGSMs due to icebreaking vessels that did not adhere to standards, divided by number of miles traveled by such vessels		H	H	M	
1.2. Enable access to icebound polar regions	Percentage of the year during which polar regions can be accessed using Coast Guard polar icebreakers		M	M	H	
1.2.1. Deploy polar-class icebreakers to polar regions as directed	Percentage of requests met for support to polar regions		H	H	H	
	Number of trips to polar regions per year		M	H	H	
	Days spent per year in polar regions		M	H	H	

Table G.3—Continued

Associated Activity, Accomplishment, or Outcome	Potential Metric	Comments on the Metric	Validity	Reliability	Feasibility	Comments on Ratings
1.3. Minimize human casualties, property damage, and environmental damage from vessels beset by ice	The number of CAGSMs caused or exacerbated by ice, normalized by the severity of the ice season and one of the following: vessel-miles, ton-miles, or dollar-miles	This includes only incidents caused or exacerbated by icy conditions. These include CAGSMs, such as vessels beset by ice.	H	M	M	
1.3.1. Rescue vessels beset by ice	Number of vessels beset by ice that the Coast Guard rescues		H	H	H	
	Number of vessels beset by ice that the Coast Guard rescues, normalized by the severity of the ice season and one of the following: vessel-miles, ton-miles, or dollar-miles		H	M	M	
1.4. Minimize human casualties and property damage from flooding due to ice jams	Number of human injuries or fatalities due to flooding from uncleared (or insufficiently cleared) ice jams normalized by severity of the season		H	H	H	
	Value of property and infrastructure damage due to flooding from uncleared (or insufficiently cleared) ice jams normalized by the severity of the season		H	M	H	
	Number of times ice-jam flooding exceeds a given threshold (height or area covered), normalized by total number of ice-jam floods to which the Coast Guard or others respond		H	H	H	

Table G.3—Continued

Associated Activity, Accomplishment, or Outcome	Potential Metric	Comments on the Metric	Validity	Reliability	Feasibility	Comments on Ratings
1.4.1. Break ice jams that pose a flooding hazard	Percentage of reported dangerous ice jams cleared, either within a threshold timespan or before a given level of flooding occurs		H	M	H	

APPENDIX H

Aids to Navigation and Waterways Management

Table H.1
Logic Model for Aids to Navigation and Waterways Management

Outcomes	Accomplishments	Activities
1. Facilitate safe and efficient use of the MTS	1.1. Reduce impediments to navigation	1.1.1. Regulate the placement and operation of bridges and offshore infrastructure over and in navigable waterways to minimize barriers to navigation
		1.1.2. Coordinate with U.S. Army Corps of Engineers and other entities to provide input on the regulation and removal of temporary and permanent potential obstructions to navigation
		1.1.3. Maintain up-to-date WAMS for planning
	1.2. Achieve high levels of mariner awareness of navigational conditions and environment	1.2.1. Regulate and plan the optimal placement and maintenance of public and private navigation infrastructure
		1.2.2. Place, maintain, and (as necessary) remove navigational infrastructure
		1.2.3. Provide maritime public with information regarding discrepancies in navigation infrastructure and other unpublished changes to maritime navigation safety
	1.3. Reduce frequency of maritime accidents resulting from special circumstances	1.3.1. Establish, maintain, and monitor limited-access areas
		1.3.2. Regulate maritime events
		1.3.3. Direct maritime traffic

Table H.2
Existing Metrics for Aids to Navigation and Waterways Management

Outcome, Accomplishment, or Activity	Existing Metric	Source Document(s)	Comments on Metric	Validity	Reliability	Feasibility	Comments on Ratings
1. Facilitate safe and efficient use of the the MTS 1.2. Achieve high levels of mariner awareness of navigational conditions and environment	CAGs	Albero, 2014, p. 12; OPAR		L	M	H	Validity: Given the availability of alternative electronic sources of information, it is doubtful that ATON shortfalls are the dominant or sole contributor to CAGs Also, these metrics could be better characterized as a rate per vessel activity Reliability is "M" because there may be different definitions of whether an event qualified as a reportable CAG.
	Five-year average number of CAGs	Michel, 2014, Enclosure 3, p. 48; Grant Thornton LLP, United States Coast Guard Waterways Management: Aids to Navigation: Independent Verification and Validation of Performance Measure Data: FY 2009 Pilot Year Review and Report of Findings, Los Angeles: Grant Thornton LLP, September 17, 2009, p. 2	This is listed in the SPD; however, it still does not include any of the marine safety inspection or licensing programs	L	M	H	

Table H.2—Continued

Outcome, Accomplishment, or Activity	Existing Metric	Source Document(s)	Comments on Metric	Validity	Reliability	Feasibility	Comments on Ratings
1.1.1. Regulate the placement and operation of bridges and offshore infrastructure over and in navigable waterways to minimize barriers to navigation	Permit application processing time	Servidio, 2013a, p. 51/55		M	H	H	Validity: Time of bridge permit application processing likely has no impact on the ultimate marine safety outcomes. If anything, slower permit processing will improve navigational safety because bridges are inherently obstructive Either way, it is a good "government efficiency" measure (management measure)
	T-H investigation processing time	Servidio, 2013a, p. 51/55		L	H	M	These are all administrative metrics, not operational ones; they bear little connection to the items they aim to capture No existing comprehensive information management system is capable of tracking this information; the program is currently building an information management system
	Drawbridge regulation processing time	Servidio, 2013a, p. 51/55		L	H	M	
	Temporary deviations processing time	Servidio, 2013a, p. 51/55		L	H	M	
	Civil penalty processing time	Servidio, 2013a, p. 51/55		L	H	M	

Table H.2—Continued

Outcome, Accomplishment, or Activity	Existing Metric	Source Document(s)	Comments on Metric	Validity	Reliability	Feasibility	Comments on Ratings
	Construction monitoring	Servidio, 2013a, p. 51/55		L	H	M	
	Bridge discrepancy processing time	Serv dio, 2013a, p. 51/55		M	H	M	
1.2.1. Regulate and plan the optimal placement and maintenance of public and private navigation infrastructure	Number of ATONs	Neffenger, 2013b, Enclosure 3, p. 41 (which was in turn taken from the Mar time Short-Range ATON Strategic Plan); FY13 USCG APR, p. 38	Removed from Michel, 2014: Reduce by 10% the number of ATON by 2014 At the end of FY13, the number was reduced by 3.1% The Coast Guard establishes, maintains, and operates more than 49,000 buoys and beacons that comprise the U.S. Visual ATON System, including almost 15,000 unlighted buoys marking the U.S. Western Rivers System	M	H	H	Validity: These are basically management and efficiency measures, not intended to be linked to activities, accomplishments, or outcomes but to resource management in the context of technological advances Numbers of items can also be a function of nonnavigational requirements, such as political considerations, or silting conditions that preclude the removal of ATONs
	Number of submarine cables	Michel, 2014 Enclosure 3, p. 51 (Which was in turn taken from the Maritime Short-Range ATON Strategic Plan)	Reduce by 75% the number of submarine cables by 2015 At the end of FY13 the number was reduced by 51.7%.	M	H	H	
	Number of fog detectors	Michel, 2014 Enclosure 3, p. 51 (which was in turn taken from the Maritime Short-Range ATON Strategic Plan)	Reduce by 50% the number of fog detectors by 2015 At the end of FY13 the number was reduced by 15.6%.	M	H	H	

Table H.2—Continued

Outcome, Accomplishment, or Activity	Existing Metric	Source Document(s)	Comments on Metric	Validity	Reliability	Feasibility	Comments on Ratings
	Number of sound signals	Michel, 2014, Enclosure 3, p. 51 (which was in turn taken from the Maritime Short-Range ATON Strategic Plan)	Reduce by 50% the number of sound signals on buoys by 2015 At the end of FY13 the number was reduced by 19%	M	H	H	
1.2.2. Place, maintain, and (as necessary) remove navigational infrastructure	Availability of maritime navigation aids (ATONs)	Neffenger, 2013b, Enclosure 2, p. 13; Servidio, 2013a, pp. 50–54; Michel, 2014, p. 42	Comment from the Michel, 2014: This overall measure tracks the percentage of time CG visual aids operate as specified	H	H	H	Age of assets, age of ATONs, and diversion of assets for other missions all influence availability in ways that do not reflect on the performance of personnel doing this mission
	Federal short-range ATON aid availability rate	Albero, 2014, p. 11; OPAR, p. 1; Servidio, 2013a, pp. 50–54; Michel, 2014 p. 42	Duplicate of SPD metric, but more specific FY 2013 actual: 98.23% FY 2014 target: 97.5%	H	H	H	
	Aid availability Cat 1	Albero, 2014, p. 12; OPAR, p. 1		H	H	H	
	Aid availability Cat 2	Albero, 2014, p. 12; OPAR, p. 1		H	H	H	
	Aid availability Cat 3	Albero, 2014, p. 12; OPAR, p. 1		H	H	H	

Table H.2—Continued

Outcome, Accomplishment, or Activity	Existing Metric	Source Document(s)	Comments on Metric	Validity	Reliability	Feasibility	Comments on Ratings
	Percentage of aids serviced on time	Albero, 2014, pp. 11, 12; OPAR; Servidio, 2013a, pp. 50–54; Michel, 2014, p. 42	FY13 Actual: Integrated Aids to Navigation System FY14 Target: 95%; NOTE: "One issue with 'percentage of aids serviced on time' is that (at least on 225' cutters) the COs [commanding officers] determine the maintenance schedules. The minimums are established in the manual, but oftentimes the cutter's schedule will be more rigorous. Therefore, it is possible that a cutter would be trying to meet their own rigorous schedule and failing, but could still manage to maintain the program standards. Just saying that the asset is meeting the bare minimum standards is not a true indicator of how effective the asset actually is."	H	H	H	
	Precision and accuracy of discrepancy report data	Michel, 2014 p. 41; Neffenger, 2013b, p. 41/54	Reduce "reasons not reported" for recording discrepancies in the Integrated Aids to Navigation System to <5%	H	M	M	Feasibility: One piece is getting the report, the other is ensuring that it is disseminated in an easily consumed format

Table H.3
Potential Metrics for Aids to Navigation and Waterways Management

Outcome, Accomplishment, or Activity	Potential Metric	Comments on the Metric	Validity	Reliability	Feasibility	Comments on Ratings
1. Facilitate safe and efficient use of the MTS	CAGSs in which ATON is determined to have been a factor in the incident, divided by vessel-miles, ton-miles, passenger-miles, and/or dollar-miles traveled	CAGSs SAR cases We are counting only CAGSs in which ATON is determined to have been a factor, rather than all CAGSs	H	M	M	A comprehensive denominator (including all boating activity) would be difficult to collect, although rough estimates are possible (perhaps from extrapolation)
	Dollars per day per mile facilitated, divided by an appropriate economic indicator, such as gross domestic product or the volume of maritime trade		M	M	M	
	"Velocity" of MTS usage (vessel-miles, ton-miles, passenger-miles, and/or dollar-miles per day) divided by U.S. gross domestic product or the total value of maritime trade	The Army Corps and Customs (CBP) already collects data that could be used to estimate MTS usage levels	M	M	M	
1.1. Reduce impediments to navigation	Number and severity of complaints of obstructions divided by vessel-miles, ton-miles, passenger-miles, and/or dollar-miles traveled		H	H	M	
	Surveys of waterway user satisfaction with respect to the waterway's navigability, divided by vessel-miles, ton-miles, passenger-miles, and/or dollar-miles traveled		H	M	L	Survey reliability is dependent on sample size
	Promptness of notification and coordination with the Army Corps to remove obstructions (measured in units of time)		H	M	H	

Table H.3—Continued

Outcome, Accomplishment, or Activity	Potential Metric	Comments on the Metric	Validity	Reliability	Feasibility	Comments on Ratings
1.1.1. Regulate the placement and operation of bridges over navigable waterways to minimize barriers to navigation	The number and severity of complaints about bridges as obstructions, divided by one or more of the following: the number of bridges, the number of vessel-miles traveled, the economic benefits of the bridges, and the cost of removal	Complaints about bridges as obstructions can include comments during the permitting process or complaints about the height or nonresponsiveness of bridge operators when openings are requested	M	M	M	
	Time to resolve complaints about openings or abandoned bridges and/or time to address comments submitted during the permitting process, divided by the total number of waterway users that are impacted by the bridge		H	M	L	Feasibility is "L" because this would be very difficult to calculate accurately (although for particular bridge permits, waterway usage estimates are likely a part of the permitting package)
	Total permit approval time, divided by the opportunity cost of not allowing sufficient time for organized citizen input/participation		H	L	M	The opportunity cost of insufficient citizen participation is a very rough guess
1.1.2. Coordinate with U.S. Army Corps of Engineers and other entities to provide input on the regulation and removal of temporary and permanent potential obstructions to navigation	U.S. Army Corps of Engineers evaluation of interactions with the Coast Guard, as measured by surveys		M	M	L	
	Audits of the review process to assess review quality		H	M	L	
1.1.3. Maintain up-to-date WAMS for planning	The percentage of regional WAMS analyses that are up to date		H	M	H	

Table H.3—Continued

Outcome, Accomplishment, or Activity	Potential Metric	Comments on the Metric	Validity	Reliability	Feasibility	Comments on Ratings
	Percentage of relevant historical MTS-related events considered during MTS improvement planning		M	M	H	
1.2. Achieve high awareness of safe and unsafe navigation areas	Maritime user feedback on the adequacy of ATON, weighted by each mariner's total usage	Feedback could be obtained by survey	H	M	L	
1.2.3. Ensure the maritime public is aware of recent changes to available navigation infrastructure and other unpublished changes to maritime navigation safety	Measure mariner awareness of last-minute changes to published MTS infrastructure and absence of obstructions (published channels), divided by vessel-miles, ton-miles, passenger-miles, and/or dollar-miles traveled	Requires a survey	H	M	L	
	Monitor mariner "consumption" of notice to mariner information using website "hits," divided by vessel-miles, ton-miles, passenger-miles, and/or dollar-miles traveled		M	M	H	
	Waterspace user awareness (based on surveys), divided by vessel-miles, ton-miles, passenger-miles, and/or dollar-miles traveled	Requires surveys	M	M	L	
	Number of CAGS-related incidents that might not have occurred if the mariner were aware of recent changes to the MTS, divided by the total number of CAGSs		H	M	M	

Table H.3—Continued

Outcome, Accomplishment, or Activity	Potential Metric	Comments on the Metric	Validity	Reliability	Feasibility	Comments on Ratings
1.3. Reduce frequency of maritime accidents resulting from special circumstances	Number of CAGSs due to special events divided by waterway usage (vessel-miles, ton-miles, passenger-miles, and dollar-miles traveled)		H	M	M	
	Number of SAR cases resulting from a poorly organized event (e.g., a swim across the bay), divided by the number of lives at risk		H	L	M	There could be disagreement about the right denominator to use
1.3.1. Establish, maintain, and monitor limited-access areas	Percentage of areas designated and approved for restrictions in accordance with published guidance, normalized by the number of locations (and events) that should have, and in fact did have, limited-access areas		M	M	M	
	Average number of days of public notice prior to the enforcement of a limited-access area		H	H	H	
	Instances in which number of days of public notice of a limited-access area prior to enforcement exceeds a minimum threshold		H	H	H	
1.3.2. Regulate maritime events	Timeliness (average number of days) for official responses (request for changes, updates, or approval) to applications for marine events		H	M	M	This could easily be high feasibility with some improved tracking tools.

Table H.3—Continued

Outcome, Accomplishment, or Activity	Potential Metric	Comments on the Metric	Validity	Reliability	Feasibility	Comments on Ratings
	Number of fatalities and significant injuries resulting from events that have received permits, divided by the number of participants and spectators; these may be grouped by event type (e.g., speed boat races, sailing activities) to reflect different risks associated with different events	Probably already close to zero in any case, but should still remain close to zero. 33 Code of Federal Regulations 100.15 says the following: "An individual or organization planning to hold a regatta or marine parade which, by its nature, circumstances or location, will introduce extra or unusual hazards to the safety of life on the navigable waters of the United States."	H	H	M	The number of spectators will always be a rough estimate
1.3.3. Direct maritime traffic	Satisfaction of Vessel Traffic Service "customers" (users) as measured by surveys and normalized by the number of Vessel Traffic Service customers per year		H	M	L	
	Number of CAGSs (including SAR) among vessels participating divided by vessel-miles, ton-miles, passenger-miles, and/or dollar-miles traveled		H	M	M	

APPENDIX I

Marine Environmental Protection

Table I.1
Logic Model for Marine Environmental Protection

Outcomes	Accomplishments	Activities
1. Decrease probability and prospective impact of spills and releases	1.1. Increase proficiency of responders and coordination with them	1.1.1. Conduct unit, responder, and industry training and awareness and maritime community outreach 1.1.2. Conduct interagency, state, local, and private-sector spill and disaster preparedness exercises, including GIUEs
	1.2. Improve response plans	1.2.1. Assist with development of industry, state, and local response plans
	1.3. Reduce risk from vessels and facilities	1.3.1. Conduct safety and MARPOL examinations of vessels and facilities
2. Mitigate effects of spills and releases through response operations	2.1. As federal on-scene coordinator, ensure that spills and releases are contained and remediated	2.1.1. Monitor and document cleanup and remediation operations, supervise contractors, and coordinate among agencies 2.1.2. Deploy pollution first responders and respond to incident
	2.2. Conduct legal investigation of spill or release incident	2.2.1. Investigate incident by documenting evidence, pollution sources, causal factors, and responsible parties, while also contributing to others' investigations 2.2.2. Prepare for legal proceedings by issuing appropriate citations and prepare case packages for prosecution
3. Monitor and communicate hazards to the public	3.1. Monitor and model flow of hazardous substances	3.1.1. Detect and track movements of hazardous substances 3.1.2. Model the anticipated movements of hazardous substances
	3.2. Inform the public to reduce exposure to hazardous substances	3.2.1. Designate evacuation and exclusion areas, based on current and anticipated hazards 3.2.2. Disseminate information regarding evacuation and exclusion areas

Table I.2
Existing Metrics for Marine Environmental Protection

Outcome, Accomplishment, or Activity	Existing Metric	Source Document(s)	Comments on the Metric	Validity	Reliability	Feasibility	Comments on Ratings
1.1.2. Conduct interagency, state, local, and private-sector spill and disaster preparedness exercises, including GIUEs	GIUE completion rate; total authorized versus number completed	Michel, 2014	Data source: MISLE	M	H	H	
	Facility/vessel GIUE compliance rate; annual number conducted vs. percentage of facilities/vessels in compliance	Michel, 2014	Data source: MISLE	M	H	H	
	Percentage exercise completion rate (conducted versus planned)	Albero, 2014	Target: 100%	M	H	H	Measure does not include quality of exercise and/or participation rate from actual responders
2. Mitigate effects of spills and releases through response operations	Percentage of mitigation of medium and major oil spills	USCG, *Marine Environmental Response Performance Plan FY2011–2016*, September 30, 2010c, Not available to the general public.	No target ID	H	H	H	
	Percentage of pollution reports that resulted in a Coast Guard response	Michel, 2014, 2014 5R Performance Review	Data source: MISLE Target: 100%	M	H	H	
	Percentage of reports that resulted in a response	Albero, 2014	Target stated as 100% However, SMEs indicated that multiple reports on the same spill are a frequent occurrence, as are spills that are too small to merit a response, so 100% may not be a desirable goal	M	H	H	

Table I.2—Continued

Outcome, Accomplishment, or Activity	Existing Metric	Source Document(s)	Comments on the Metric	Validity	Reliability	Feasibility	Comments on Ratings
	Percentage of spilled material recovered for medium and major oil spills	USCG, 2010c	No target ID	H	H	H	
2.1.1. Monitor and document cleanup and remediation operations, supervise contractors, and coordinate among agencies 2.2.1. Investigate incident by documenting evidence, pollution sources, causal factors, and responsible parties, while also contributing to others' investigations	Percentage of AARs submitted	USCG, 2010c	Target 25% for real-world events from units seeking exercise credit	M	H	H	An AAR is documentation, but not part of the case package
N/A	MEP efficiency ratio	Marine Safety Mission Performance Plan FY11–16	Target 1.00 (ratio of five-year average annual numbers of spills and discharges, prior period to current period, divided by five-year operating expense for the program)	L	M	H	

Table I.3
Potential Metrics for Marine Environmental Protection

Associated Activity, Accomplishment, or Outcome	Potential Metric	Comments on the Metric	Validity	Reliability	Feasibility	Comments on Ratings
1. Decrease probability and prospective impact of spills and releases	Estimated reduction in effects of spills and releases due to Coast Guard actions	Requires risk modeling	M	L	L	
1.1. Increase proficiency of responders and coordination with them	Estimated percentage of relevant individuals achieving a given training standard	Likely requires quizzes and/or extrapolation from small samples	M	L	L	
	Percentage of entities exceeding a given performance level in exercises	Involves considerable subjectivity	M	L	L	
1.1.1. Conduct unit, responder, and industry training and awareness and maritime community outreach	Number of training events held, categorized by type of event		M	H	H	
	Number of outreach and awareness events held, categorized by type of event		M	H	H	
	Percentage of training audience engaged by one or more training events	Would be based on surveys and estimates	H	L	L	
	Percentage of outreach and awareness audience effectively reached	Would be based on surveys and estimates	H	L	L	
1.1.2. Conduct interagency, state, local, and private-sector spill and disaster preparedness exercises, including GIUEs	Number of exercises conducted, categorized by type of exercise		H	H	H	
	Percentage of relevant entities participating in at least one exercise		H	H	H	
1.2. Improve response plans	Percentage of response plans that meet a given Coast Guard standard		M	M	M	

Table I.3—Continued

Associated Activity, Accomplishment, or Outcome	Potential Metric	Comments on the Metric	Validity	Reliability	Feasibility	Comments on Ratings
1.2.1. Assist with development of industry, state, and local response plans	Number of plans to which the Coast Guard contributes	The scale of contribution can obviously vary greatly, from collaborative composition to cursory review, but we are making it a binary distinction (whether the Coast Guard contributed or not)	M	H	H	
	Percentage of relevant plans to which the Coast Guard contributes	The scale of contribution can obviously vary greatly, from collaborative composition to cursory review, but we are making it a binary distinction (whether the Coast Guard contributed or not)	M	H	H	
1.3. Reduce risk from vessels and facilities	Percentage of vessels and facilities not achieving a given level of safety and MARPOL compliance		M	M	M	
	Estimated probability of spills and releases of given magnitudes	Requires modeling and estimation	M	L	L	
1.3.1. Conduct safety and MARPOL examinations of vessels and facilities	Number of safety and MARPOL examinations of vessels the Coast Guard conducts		M	H	H	
	Number of safety and MARPOL examinations of facilities the Coast Guard conducts		M	H	H	
	Percentage of vessels on which the Coast Guard conducts safety and/or MARPOL examinations		M	H	H	
	Percentage of facilities on which the Coast Guard conducts safety and/or MARPOL examinations		M	H	H	

Table I.3—Continued

Associated Activity, Accomplishment, or Outcome	Potential Metric	Comments on the Metric	Validity	Reliability	Feasibility	Comments on Ratings
2. Mitigate effects of spills and releases through response operations	None needed—covered by existing metrics					
2.1. As federal on-scene coordinator, ensure that spills and releases are contained and remediated	Economic impact of spill, including economic costs of environmental damage	Will be influenced by many factors beyond the Coast Guard's control	L	L	L	
2.1.1. Monitor and document cleanup and remediation operations, supervise contractors, and coordinate among agencies	Percentage of material recovered	Will be influenced by many factors beyond the Coast Guard's control	L	M	M	
	Number of safety incidents arising during course of response	This will be influenced by many factors beyond the Coast Guard's control	L	M	M	
2.1.2. Deploy pollution first responders and respond to incident	Percentage of instances in which first responders arrive within a designated time frame		M	H	H	
2.2. Conduct legal investigation of spill or release incident	Percentage of incidents in which prosecution is pursued and successful	This will be influenced by many factors beyond the Coast Guard's control	L	H	H	
2.2.1. Investigate incident by documenting evidence, pollution sources, causal factors, and responsible parties, while also contributing to others' investigations	Percentage of incidents in which the Coast Guard completes its investigation	This will likely be 100%	M	H	H	
	Percentage of incidents in which pollution sources are determined	These depend on many factors beyond the Coast Guard's control	L	H	H	
	Percentage of incidents in which causal factors are determined		L	H	H	
	Percentage of incidents in which the Coast Guard contributes to other agencies' investigations		L	M	H	

Table I.3—Continued

Associated Activity, Accomplishment, or Outcome	Potential Metric	Comments on the Metric	Validity	Reliability	Feasibility	Comments on Ratings
2.2.2. Prepare for legal proceedings by issuing appropriate citations and prepare case packages for prosecution	Number of instances in which citations are issued, divided by the number in which they should be		M	M	H	
	Number of instances in which case packages are completed, divided by the number in which they should be		M	M	H	
	Percentage of successful prosecutions	Many factors other than the case package contribute to this	L	H	M	
3. Monitor and communicate hazards to the public	Number of individuals exceeding various threshold levels of exposure (e.g., high, medium, low)	Will be influenced by many factors beyond the Coast Guard's control, and is hard to measure accurately	M	L	L	
	Estimated reduction in number of individuals exceeding threshold levels of exposure due to Coast Guard actions	Will involve very rough estimates	H	L	L	
3.1. Monitor and model flow of hazardous substances	External ratings of Coast Guard modeling effectiveness, on a scale of 0 to 10	Would require a survey of SMEs who have worked with the Coast Guard	M	M	L	
3.1.1. Detect and track movements of hazardous substances	Resolution of detection and tracking of hazardous substances (e.g., 1 m, 10 m, 100 m)		L	M	M	
3.1.2. Model the anticipated movements of hazardous substances	Retrospective accuracy of modeling, measured in terms of percentage overlap of hazardous areas at any given time		M	M	M	

Table I.3—Continued

Associated Activity, Accomplishment, or Outcome	Potential Metric	Comments on the Metric	Validity	Reliability	Feasibility	Comments on Ratings
3.2. Inform the public to reduce exposure to hazardous substances	Number of individuals leaving or avoiding affected areas	Heavily dependent on factors beyond the Coast Guard's control and involves a large margin of error; how many people would have been expected to be in an affected area if the event had not happened may also be ambiguous	M	L	L	
3.2.1. Designate evacuation and exclusion areas, based on current and anticipated hazards	Extent of evacuation and exclusion areas, in square miles		L	M	M	
	Retrospective assessment of percentage of evacuation or exclusion areas that coincided with actual hazard		M	M	M	
3.2.2. Disseminate information regarding evacuation and exclusion areas	Number of communications channels used (radio, Internet, etc.)		L	H	H	
	Estimated percentage of the relevant public reached within a given (or required) time frame	Postevent surveys would be required	H	M	L	

APPENDIX J

Living Marine Resources

Table J.1
Logic Model for Living Marine Resources

Outcomes	Accomplishments	Activities
1. Prevent illegal fishing activities and other threats to marine species	1.1. Communicate and work with both other agencies and the general public	1.1.1. Continue and expand partnerships with other agencies
		1.1.2. Share information with the public
	1.2. Enforce regulations	1.2.1. Maintain effective presence to achieve awareness and deter prospective violations
		1.2.2. Monitor for and detect overt LMR violations (e.g., based on location, timing, activity, gear)
		1.2.3. Detect and deter less-visible LMR violations by conducting random and/or targeted boardings
		1.2.4. Respond to LMR violations—document, disrupt, intercept, interdict, board and apprehend, and process, as appropriate
		1.2.5. Cite violators and prepare cases for prosecution, as appropriate
	1.3. Contribute to conservation efforts	1.3.1. Monitor, collect, analyze, and disseminate information on specific marine species
	1.4. Ensure Coast Guard compliance with laws and regulations for LMR	1.4.1. Engage with relevant Coast Guard units and programs to ensure knowledge of and adherence to laws and regulations

Table J.2
Existing Metrics for Living Marine Resources

Outcome, Accomplishment, or Activity	Existing Metric	Source Document(s)	Comments on the Metric	Validity	Reliability	Feasibility	Comments on Ratings
1.2. Enforce regulations	Percentage of high- and low-precedence fisheries with satisfactory levels of enforcement	OPAR	Reported by district	H	H	H	
	Level of effective enforcement			H	M	M	
1.2.1. Maintain effective presence to achieve awareness and deter prospective violations	District compliance with reporting number of active domestic fishing vessels, by fishery	Michel, 2014, p. 18	Reported by district	H	M	M	
1.2.3. Detect and deter less-visible LMR violations by conducting random and/or targeted boardings	Percentage boarding rate for U.S. fishing fleet operating in high-precedence fisheries	Michel, 2014, p. 18; USCG, 2014a, p. 9		H	M	H	Denominator may not be precisely known in some areas
	Percentage boarding rate for U.S. fishing fleet operating in low-precedence fisheries	Michel, 2014, p. 18; USCG, 2014a, p. 9		H	M	H	Denominator may not be precisely known in some areas
	Surveil high-threat areas to detect 80% of significant violations	USCG, 2014a, pp. 8–9		M	M	M	Denominator may not be precisely known in some areas
	Surveil low-threat areas to detect 20% of all significant violations	USCG, 2014a, p. 9		M	M	M	Denominator may not be precisely known in some areas
1.2.4. Respond to LMR violations—document, disrupt, intercept, interdict, board and apprehend, and process, as appropriate	Response rate to known significant violations in progress	USCG, 2014a, p. 8		M	M	M	

Table J.2—Continued

Outcome, Accomplishment, or Activity	Existing Metric	Source Document(s)	Comments on the Metric	Validity	Reliability	Feasibility	Comments on Ratings
1.2.5. Cite violators and prepare cases for prosecution as appropriate	Document every significant violation of domestic LMR laws and regulations discovered	USCG, 2014a, p. 9		M	M	M	
1.3. Contribute to conservation efforts	Conservation assistance requests	OPAR		H	H	H	
1.3.1. Monitor, collect, analyze, and disseminate information on specific marine species	Percentage of requests for LMR conservation assistance met	OPAR		H	H	H	
	Number of marine protected species pulse operations	OPAR	At least one required per year per district	M	H	H	
	Number of collaborative marine protected-species outreach and education campaigns	OPAR	At least one required per year per district	M	H	H	

Table J.3
Potential Metrics for Living Marine Resources

Associated Activity, Accomplishment, or Outcome	Potential Metric	Comments on the Metric	Validity	Reliability	Feasibility	Comments on Ratings
1. Prevent illegal fishing activities and other threats to marine species	NOAA ratings of the health of specific fisheries and protected species	The health of fisheries and protected species is highly dependent on factors outside the Coast Guard's control, but is the ultimate end of this mission	L	M	M	
1.1. Communicate and work with both other agencies and the general public	Extent to which other agencies view the Coast Guard as collaborating well with them, on a 0–10 scale	Would require surveys of relevant stakeholders	M	M	L	
	Estimated degree to wh ch public outreach and campaigns affect fishing behaviors	Would require surveys, with some uncertainty about the degree to which survey answers reflected actual behavior	M	M	L	
1.1.1. Continue and expand partnerships with other agencies	Number of high-level Coast Guard meetings with interagency and international partners about LMR		M	H	H	
	Number of interagency agreements (e.g., MOA, MOU), as s percentage of the desired ones		M	H	H	
	Number of bilateral or multilateral agreements, as a percentage of the desired ones		M	H	H	
	Percentage of interagency bodies and centers dealing with LMR in which the Coast Guard has appropriate representation		M	H	H	

Table J.3—Continued

Associated Activity, Accomplishment, or Outcome	Potential Metric	Comments on the Metric	Validity	Reliability	Feasibility	Comments on Ratings
	Number of full-time Coast Guard liaison personnel addressing LMR at other agencies or interagency centers or bodies, divided by the desirable number of liaison personnel (or divided by the number of agencies plus the number of bodies or centers)		M	H	H	
	Number of other agencies' and interagency liaison personnel addressing LMR at relevant Coast Guard commands		M	H	H	
	Number of Coast Guard shipriders aboard other agencies' vessels conducting at least some LMR activities		M	H	H	
	Number of other agencies' shipriders aboard Coast Guard vessels conducting at least some LMR activities		M	H	H	
	Number of multiagency LMR exercises in which the Coast Guard participates		M	H	H	
	Percentage of multiagency LMR exercises in which the Coast Guard participates		M	H	H	
	Number of LMR operations conducted in cooperation with other agencies		M	H	H	
	Percentage of requests for information responded to		M	H	H	
1.1.2. Share information with the public	Number of public-affairs campaigns being conducted		M	H	H	

Table J.3—Continued

Associated Activity, Accomplishment, or Outcome	Potential Metric	Comments on the Metric	Validity	Reliability	Feasibility	Comments on Ratings
	Forms of communication being used by public-affairs campaigns (e.g., radio, television, Internet)		M	H	H	
	Estimated percentage of relevant audiences (professional and/or recreational anglers) being reached by public-affairs campaigns	Would require surveys for accuracy	M	M	L	
	Estimated percentage of relevant audiences (professional and/or recreational anglers) learning key information from public-affairs campaigns	Would require quizzes	M	M	L	
1.2. Enforce regulations	Percentage compliance, by fishery and regulation	Would be assessed based on random sampling	H	M	M	
1.2.1. Maintain effective presence to achieve awareness and deter prospective violations	Percentage of the time that vessels in a given fishery can expect to have a Coast Guard platform within periodic visual range	In this case, periodic means "frequently enough to deter violations"; the time span involved depends on the fishery and typical violations associated with it	M	L	M	
	Estimated percentage of fishing vessels that the Coast Guard knows about at any given time, averaged over time	Obviously a "known unknown" but can be estimated via extrapolation and modeling	M	L	L	
1.2.2. Monitor for and detect overt LMR violations (e.g., based on location, timing, activity, gear)	Estimated probability of detection of overt LMR violations	Requires modeling of both Coast Guard and violator behavior	M	L	L	
	Number of overt LMR violations detected, by type and fishery		M	H	H	

Table J.3—Continued

Associated Activity, Accomplishment, or Outcome	Potential Metric	Comments on the Metric	Validity	Reliability	Feasibility	Comments on Ratings
	Estimated percentage of vessels involved in fishery committing overt violations	Requires extrapolation and modeling	M	L	L	
	Estimated number of vessels involved in fishery committing overt violations	Requires extrapolation and modeling	M	L	L	
1.2.3. Detect and deter less-visible LMR violations by conducting random and/or targeted boardings	Random boardings as a percentage of all vessels in fishery		H	M	M	
	Targeted boardings as a percentage of all vessels in fishery		H	M	M	
1.2.4. Respond to LMR violations—document, disrupt, intercept, interdict, board and apprehend, and process, as appropriate	Percentage of detections leading to interceptions		M	M	M	
	Percentage of detections leading to interdictions		M	M	M	
	Percentage of interdictions leading to successful imposition of penalties		M	M	M	
	Percentage of detected violations leading to imposition of penalties		M	M	M	
1.2.5. Cite violators and prepare cases for prosecution as appropriate	Percentage of times in which evidence was unable to be used in case		H	H	M	
	Percentage of desired case packages completed		M	H	H	
	Percentage of cases in which prosecution is pursued and is successful	Would depend on many factors outside the Coast Guard's control	L	H	H	

Table J.3—Continued

Associated Activity, Accomplishment, or Outcome	Potential Metric	Comments on the Metric	Validity	Reliability	Feasibility	Comments on Ratings
1.3. Contribute to conservation efforts	External ratings (0–10) of Coast Guard contributions to these conservation efforts	Surveys would be taken by NOAA personnel and other stakeholders	M	M	L	
1.3.1. Monitor, collect, analyze, and disseminate information on specific marine species	None needed—covered by existing metrics					
1.4. Ensure Coast Guard compliance with laws and regulations for LMR	Number of incidents of Coast Guard noncompliance with laws or regulations for LMR		H	M	M	
1.4.1. Engage with relevant Coast Guard units and programs to ensure knowledge of and adherence to laws and regulations	Estimated percentage of relevant Coast Guard audience reached via engagement	Would require a survey for accuracy	L	M	L	
	Percentage of relevant Coast Guard audience demonstrating requisite knowledge	Would require a quiz	H	M	L	

APPENDIX K

Other Law Enforcement

Table K.1
Logic Model for Other Law Enforcement

Outcomes	Accomplishments	Activities
1. Enforce U.S. EEZ	1.1. Deter illegal foreign fishing in the U.S. EEZ	1.1.1. Maintain effective presence to deter prospective EEZ violations
		1.1.2. Publicize the risks and costs of committing EEZ violations
	1.2. Counter illegal foreign fishing in the U.S. EEZ	1.2.1. Monitor for and detect foreign ships illegally fishing in the U.S. EEZ
		1.2.2. Intercept and interdict FFVs
		1.2.3. Board, apprehend, collect evidence on, and cite FFVs
		1.2.4. Prepare case packages for prosecution, as appropriate
2. Enforce adherence to international fishing regulations	2.1. Prevent violations of international fishing regulations	2.1.1. Continue and expand effective partnerships to improve partner capabilities and information sharing
		2.1.2. Maintain effective presence to achieve awareness and to deter prospective violations of international regulations by demonstrating the ability to enforce them
		2.1.3. Document and respond to violations of international fishing regulations—disrupt, intercept, interdict, board, apprehend, cite, and/or prepare case packages for prosecution, as appropriate

Table K.2
Existing Metrics for Other Law Enforcement

Outcome, Accomplishment, or Activity	Existing Metric	Source Document(s)	Comments on the Metric	Validity	Reliability	Feasibility	Comments on Ratings
1.1. Deter illegal foreign fishing in EEZ	Number of instances of FFVs operating in U.S. EEZ	Michel, 2014, p. 18	Data source is monthly LMR summary/MISLE	H	L	L	We know how many were detected, but estimating the number of undetected FFVs requires modeling
1.2.1. Monitor for and detect foreign ships illegally fishing in the U.S. EEZ	Percentage of all significant violations to high-threat areas of EEZ detected	USCG, 2014a, pp. 8–9	Numbers are based on threat assessments and being able to surveil high/low threat areas sufficiently to detect percentages of violations	H	L	L	Denominator is inherently challenging to assess for some districts
	Percentage of all significant violations to low-threat areas of EEZ detected	USCG, 2014a, pp. 8–9	Numbers are based on threat assessments and being able to surveil high/low threat areas sufficiently to detect percentages of violations	H	L	L	
1.2.2. Intercept and interdict FFVs	Percentage interception rate for detected FFVs operating in U.S. EEZ	Michel, 2014, p. 18	Data source is monthly LMR summary/MISLE	M	H	H	Previously included non–Coast Guard interceptions, but information flow from other organizations is sometimes incomplete
	Percentage interdiction rate for detected FFVs operating in U.S. EEZ	Michel, 2014, p. 18	Data source is monthly LMR summary/MISLE	M	H	H	
	Interception rate of known suspects	USCG, 2014a, p. 9		M	H	H	By definition, the suspects intercepted are known. This makes reliability and feasibility high, but validity medium, since unknown suspects are not included

Table K.3
Potential Metrics for Other Law Enforcement

Associated Activity, Accomplishment, or Outcome	Potential Metric	Comments on the Metric	Validity	Reliability	Feasibility	Comments on Ratings
1. Enforce U.S. EEZ	Estimated economic losses due to FFVs fishing in EEZ		M	L	L	
	Number of species with significantly diminished numbers due to FFV fishing in EEZ		M	M	L	
1.1. Deter illegal foreign fishing in EEZ	Estimated net cost/benefit for foreign vessels to fish in EEZ, by fishery	This would entail estimating the benefits of fishing in the EEZ (value of the catch minus costs to get there, compared with the opportunity costs elsewhere), subtracting the probabilities of detection and interdiction multiplied by their consequences	M	L	M	
1.1.1. Maintain effective presence to deter prospective EEZ violations	Percentage of relevant high-threat fishing areas and times in which the Coast Guard has visible assets present		M	M	M	
	Percentage of relevant low-threat fishing areas and times in which the Coast Guard has visible assets present		M	M	M	
	Number of known EEZ violations in a given geographic area and timespan	A challenge in using this metric is that improving Coast Guard situational awareness (enabling detection of violations) makes the numbers look worse	M	H	H	
1.1.2. Publicize the risks and costs of committing EEZ violations	Number of public-affairs campaigns to deter EEZ violations		M	M	M	

Table K.3—Continued

Associated Activity, Accomplishment, or Outcome	Potential Metric	Comments on the Metric	Validity	Reliability	Feasibility	Comments on Ratings
1.2. Counter illegal foreign fishing in the U.S. EEZ	Percentage of detections resulting in successful imposition of penalties		M	M	M	
	Percentage of interceptions resulting in successful imposition of penalties		M	M	M	
	Percentage of interdictions resulting in successful imposition of penalties		M	M	M	
	Percentage of boardings in which nonviable catch is found		M	M	M	
	Total quantity of nonviable catch in tons for a given location and timespan		M	M	M	
1.2.1. Monitor for and detect foreign ships illegally fishing in the U.S. EEZ	Percentage of relevant high-threat fishing areas and times that the Coast Guard is monitoring for FFVs		M	M	M	
	Percentage of relevant low-threat fishing areas and times that the Coast Guard is monitoring for FFVs		M	M	M	
1.2.2. Intercept and interdict FFVs	Percentage of detections resulting in interceptions		M	M	M	
	Percentage of interceptions resulting in interdictions		M	M	M	
1.2.3. Board, apprehend, collect evidence on, and cite FFVs	Percentage of boardings resulting in citations		M	M	M	
1.2.4. Prepare case packages for prosecution, as appropriate	Percentage of desired case packages completed		M	H	H	

Table K.3—Continued

Associated Activity, Accomplishment, or Outcome	Potential Metric	Comments on the Metric	Validity	Reliability	Feasibility	Comments on Ratings
	Percentage of cases in which prosecution was pursued and was successful	Much of this is beyond the Coast Guard's control.	M	H	H	
2. Enforce adherence to international fishing regulations	Estimated economic damage due to fishing in violation of international regulations		M	L	L	
	Number of species with significantly diminished numbers due to violation of international regulations		M	L	L	
2.1. Prevent violations of international fishing regulations	Estimated frequency of violations of specific international fishing violations		M	L	L	This is a "known unknown"—estimates for compliance over a large fraction of the globe will be very rough
	Estimated impact of Coast Guard actions on frequency of violations of specific international fishing violations		H	L	L	This requires modeling of adversary behavior, and will likely be a very rough estimate
2.1.1. Continue and expand effective partnerships to improve partner capabilities and information sharing	Number of high-level Coast Guard meetings with interagency and international partners regarding international fishing regulations		M	H	H	
	Number of interagency agreements (e.g., MOA, MOU) regarding international fishing regulations divided by desired number		M	H	H	
	Number of bilateral or multilateral agreements regarding international fishing regulations divided by desired number		M	H	H	

Table K.3—Continued

Associated Activity, Accomplishment, or Outcome	Potential Metric	Comments on the Metric	Validity	Reliability	Feasibility	Comments on Ratings
	Number of full-time Coast Guard liaison personnel working with other countries regarding international fishing regulations		M	H	H	
	Number of other nations' liaison personnel at relevant Coast Guard commands dealing with international fishing regulations		M	H	H	
	Number of Coast Guard shipriders aboard other nations' vessels dealing with international fishing regulations		M	H	H	
	Number of other nations' shipriders aboard Coast Guard vessels dealing with international fishing regulations		M	H	H	
	Number of bilateral or multilateral training events or exercises regarding fisheries to which the Coast Guard contributes		M	H	H	
	Percentage of annual bilateral or multilateral training events or exercises regarding fisheries to which the Coast Guard contributes		M	M	H	
	Number of OLE operations conducted in cooperation with other nations		M	M	H	

Table K.3—Continued

Associated Activity, Accomplishment, or Outcome	Potential Metric	Comments on the Metric	Validity	Reliability	Feasibility	Comments on Ratings
2.1.2. Maintain effective presence to achieve awareness and to deter prospective violations of international regulations by demonstrating the ability to enforce them	Percentage of relevant fishery times and places at which Coast Guard has visible presence	Likely to be extremely low because the denominator encompasses most of the world's waters	H	M	M	
	Percentage of relevant fishery times and places that the Coast Guard is effectively monitoring	Likely to be extremely low because the denominator encompasses most of the world's waters	H	M	M	
2.1.3. Document and respond to violations of international fishing regulations—disrupt, intercept, interdict, board, apprehend, cite, and/or prepare cases for prosecution, as appropriate	Percentage of detections resulting in citations		H	M	M	
	Percentage of detections resulting in interceptions		M	H	H	
	Percentage of interceptions resulting in interdictions		M	H	H	
	Percentage of interdictions leading to successful prosecutions		H	M	M	
	Percentage of desired case packages completed		H	H	H	

Abbreviations

AAR	after-action report
AMSC	Area Maritime Security Committee
AMVER	Automatic Maritime Vessel Emergency Response
ATON	aids to navigation
AUF	airborne use of force
C2	command and control
CAGs	collisions, allisions, and groundings
CAGSs	collisions, allisions, groundings, and search and rescue cases
CAGSMs	collisions, allisions, groundings, search and rescue cases, and mishaps
CBP	Customs and Border Protection
CCDB	Consolidated Counterdrug Database
CFV	commercial fishing vessel
CGBI	Coast Guard Business Intelligence
CG-MLE	Coast Guard Office of Law Enforcement
CG-ODO	Coast Guard Office of Counterterrorism & Defense Operations Policy
CG-RACE	Coast Guard Resource and Capabilities Evaluation System
COMDTINST	commandant instruction
DHS	Department of Homeland Security
DoD	Department of Defense
DoJ	Department of Justice
DOMICE	Domestic Icebreaking
DTO	drug-trafficking organization
EEZ	exclusive economic zone

FFV	foreign fishing vessel
FY	fiscal year
GIUE	government-initiated unannounced exercise
H	high
ICC	Intelligence Coordination Center
ISPS	international ship and port security
ISR	intelligence, surveillance, and reconnaissance
JIATF-S	Joint Interagency Task Force--South
K9	canine
L	low
LANTAREA	U.S. Coast Guard Atlantic Area
LANT-OPD	Atlantic Area Operational Planning Direction
LEDET	law enforcement detachment team
LMR	Living Marine Resources
M	medium
MARPOL	International Convention for the Prevention of Pollution by Ships
MEP	Maritime Environmental Protection
MIFC	Maritime Intelligence Fusion Center
MISLE	Maritime Information for Safety and Law Enforcement
MOA	memorandum of agreement
MOTR	Maritime Operational Threat Response
MOU	memorandum of understanding
MSRAM	Maritime Security Risk Analysis Model
MSRO	Maritime Security and Response Operations
MSRT	Maritime Security Response Team
MSST	Maritime Safety and Security Team
MTS	Marine Transportation System
MTSA	Maritime Transportation Security Act of 2002
NOAA	National Oceanographic and Atmospheric Administration
OAM	Office of Air and Marine

OCO	overseas contingency operations
OLE	Other Law Enforcement
OPAR	Operational Planning Assessment Report
OPD	Operational Planning Direction
PPD	Presidential Policy Directive
PSU	Port Security Unit
PWCS	Ports, Waterways, and Coastal Security
RAID	redeployment assistance and inspection detachment
RWAI	rotary-wing air intercept operations
SAR	search and rescue
SME	subject-matter expert
SORTS	Status of Readiness and Training System
SPD	Strategic Planning Direction (memorandums)
SRU	search and rescue unit
SURPIC	surface picture
TCO	transnational criminal organization
TPS	transit protection support
TSC	theater security cooperation
TWIC	transportation worker identification credentialing
USCG	U.S. Coast Guard
USN	U.S. Navy
USSOUTHCOM	U.S. Southern Command
VBSS	visit, board, search, and seizure
WAMS	Waterway Analysis and Management System
WPB	110-foot Patrol Boat
WWM	waterway management

Bibliography

Abel, D. B., "FY 2014 First District Operational Planning Direction (OPD)," memorandum, Boston, Mass.: U.S. Coast Guard First District, December 23, 2013, Not available to the general public.

Albero, Frank, "Info Paper: FY13 Operational Performance," memorandum, Portsmouth, Va.: U.S. Coast Guard Atlantic Area, November 5, 2013, Not available to the general public.

———, "Info Paper: Review of LANTAREA Operational Planning Direction (OPD) Performance Measures," memorandum, Portsmouth, Va.: U.S. Coast Guard Atlantic Area, April 8, 2014, Not available to the general public.

Baisuck, A., and W. A. Wallace, "A Framework for Analyzing Marine Accidents," *Marine Technology Society Journal*, Vol. 13, No. 5, October–November 1979, pp. 8–14.

Brooks, A. E., "Seventeenth District Performance Plan," Juneau, Alaska, November 30, 2006.

Brown, Peter J., *United States Coast Guard Maritime Response Program Performance Plan, Fiscal Years 2014–2019*, Washington, D.C.: U.S. Coast Guard, August 2013, Not available to the general public.

Coast Guard Publication 1, *Doctrine for the U.S. Coast Guard*, Washington, D.C.: U.S. Coast Guard, February 2014.

——— 2-0, *Intelligence*, Washington, D.C.: U.S. Coast Guard, May 2010.

——— 3-0, *Operations*, Washington, D.C.: U.S. Coast Guard, February 2012.

——— 7-0, *Capability Management*, Washington, D.C.: U.S. Coast Guard, May 2013.

Code of Federal Regulations, Title 33, Navigation and Navigable Waters.

Code of Federal Regulations, Title 33, Navigation and Navigable Waters, Section 100.15, Submission of Application.

COMDTINST—*See* Commandant Instruction.

Commandant Instruction 3120.4A, "Coast Guard Standard Operational Planning Process/Global Force Management," Washington, D.C.: U.S. Coast Guard, December 10, 2009.

Commandant Instruction Manual 16130.2F, "The United States Coast Guard Addendum to the United States National Search and Rescue Supplement (NSS) to the International Aeronautical and Maritime Search and Rescue Manual (IAMSAR)," Washington, D.C.: U.S. Coast Guard, January 7, 2013.

——— 5000.3B, *United States Coast Guard Regulations*, 1992. As of September 28, 2015: http://www.uscg.mil/directives/cim/5000-5999/CIM_5000_3B.pdf

——— 16600.6, *Maritime Security Response Operations Manual*, 2011, Not available to the general public.

Crowley, John E., Jr., "Memorandum of Agreement Between the United States Coast Guard and the Bay Mills Indian Community, the Sault Ste. Marie Tribe of Chippewa Indians, the Grand Traverse Band of Ottawa and Chippewa Indians, Little River Band of Ottawa Indians," Cleveland, Ohio, October 2007.

Davis, Paul K., "Objectives, Risks, and Analysis in Strategic Planning: End-of-Workshop Observations," briefing presented at the MORS/SARMA Workshop, Alexandria, Va., October 14–17, 2014.

Downer, Kevin, and Timothy Hughes, "Alien Migrant Interdiction Operations (AMIO) Game Theory Scoping Study," New London, Conn.: U.S. Coast Guard Research and Development Center, May 27, 2014, Not available to the general public.

Energetics, *United States Coast Guard Waterways Management: Aids to Navigation: Post-Assessment Improvement Actions: Review and Report of Findings, draft*, Columbia, Md., May 2011.

———, *United States Coast Guard Ports, Waterways, and Coastal Security: Independent Verification and Validation of Performance Measure Data: FY 2012 Review and Report of Findings,* Columbia, Md., February 2012.

———, *United States Coast Guard Maritime Prevention Program: Independent Verification and Validation of Performance Measure Data: FY 2013 Review and Report of Findings*, Columbia, Md., April 2013.

———, *Office of Intelligence & Analysis DHS Intelligence Enterprise: Independent Verification and Validation of Performance Measure Data: FY 2014 Review and Report of Findings*, draft, Columbia, Md., February 2014.

Fifth Coast Guard District Prevention Division, Waterways Management Branch, "Executive Overview: Mission Management System," undated.

Fifth Coast Guard District Waterways Management Branch, "Domestic Icebreaking Operations: Ice Season 2014-2015," briefing slides, Portsmouth, Va., undated.

FY 2015 SPD—*See* Michel, 2014.

Grant Thornton LLP, *United States Coast Guard Waterways Management: Aids to Navigation: Independent Verification and Validation of Performance Measure Data: FY 2009 Pilot Year Review and Report of Findings*, Los Angeles: Grant Thornton LLP, September 17, 2009.

Greenfield, Victoria A., Valerie L. Williams, and Elisa Eiseman, *Using Logic Models for Strategic Planning and Evaluation: Application to the National Center for Injury Prevention and Control*, Santa Monica, Calif.: RAND Corporation, TR-370-NCIPC, 2006. As of August 5, 2015:
http://www.rand.org/pubs/technical_reports/TR370.html

Homeland Security Studies and Analysis Institute, "Methodology for Improving the U.S. Coast Guard Operational Risk Assessment Model (ORAM)," May 2011.

———, "Operational Risk Assessment Model (ORAM) Optimization Modeling," April 2014.

———, "Operational Risk Assessment Model (ORAM): Progress Has Been Made But More Needs to Be Done to Provide Actionable Information," July 2013.

Johnson, Jeh Charles, "Southern Border and Approaches Campaign," memorandum distributed to the heads of Department of Homeland Security components, November 20, 2014.

Joint Chiefs of Staff, *The National Military Strategy of the United States of America 2015*, Washington, D.C., June 2015. As of September 4, 2015:
http://www.jcs.mil/Portals/36/Documents/Publications/2015_National_Military_Strategy.pdf

Korn, J. H., "District Seven FY14 Operational Planning Direction (OPD)," memorandum distributed to Coast Guard sectors and air stations, Miami, Fla.: U.S. Coast Guard Seventh District, December 27, 2013.

Kostecki, K., L. Fisher, and M. Farrell, "LANTAREA Api Strategic Business Plan: FY 2007," paper, Washington, D.C.: U.S. Coast Guard, October 27, 2006.

Lee, W. D., "FY2015 LANTAREA Operational Planning Direction," memorandum, Portsmouth, Va.: U.S. Coast Guard Atlantic Area, September 19, 2014, Not available to the general public.

Lescher, William, "Navy FY15 Budget: Preserving Presence and Warfighting Capability," Navy Live blog, March 4, 2014. As of August 26, 2015:
http://navylive.dodlive.mil/2014/03/04/navy-fy15-budget-preserving-presence-and-warfighting-capability/

McLaughlin, John A., and G. B. Jordan, "Logic Models: A Tool for Telling Your Program's Performance Story," *Evaluation and Program Planning*, Vol. 22, No. 1, 1999, pp. 65–72.

———, "Using Logic Models," in Joseph S. Wholey, Harry P. Hatry, and Kathryn E. Newcomer, eds., *Handbook of Practical Program Evaluation*, San Francisco: Jossey-Bass, 2004, pp. 7–32.

Michel, Charles D., "Fiscal Year 2015 Strategic Planning Direction," memorandum, Washington, D.C.: U.S. Coast Guard, August 13, 2014, Not available to the general public.

Midgette, F. M., "D9 FY2014 Operational Planning Direction," memorandum, Cleveland, Ohio: U.S. Coast Guard Ninth District, February 26, 2014, Not available to the general public.

Neffenger, Peter V., "DCO Strategic Management Responsibilities for Operating Programs," memorandum, Washington, D.C.: U.S. Coast Guard, February 21, 2013a, Not available to the general public.

———, "Fiscal Year 2014 Strategic Planning Direction (SPD)," memorandum, Washington, D.C.: U.S. Coast Guard, September 13, 2013b, Not available to the general public.

Office of National Drug Control Policy, *National Southwest Border Counternarcotics Strategy*, 2013.

Office of Performance Management and Assessment (DCO-81), *U.S. Coast Guard Counter Drug Mission Analysis Report*, Washington, D.C.: Office of Performance Management and Assessment (DCO-81), September 30, 2013, Not available to the general public.

O'Rourke, Ronald, "Navy Ford (CVN-78) Class Aircraft Carrier Program: Background and Issues for Congress," Washington, D.C.: Congressional Research Service, RS20643, October 22, 2013.

Palin, Philip, Steven Hall, Ted Lewis, and Craig Baldwin, "Deterrence and the United States Coast Guard: Enhancing Current Practice with Performance Measures," Monterey, Calif.: Center for Homeland Defense and Security, Naval Postgraduate School, March 2012.

Parker, R. C., "FY14 Operational Planning Direction," memorandum, Portsmouth, Va.: U.S. Coast Guard Atlantic Area, November 27, 2013, Not available to the general public.

Peoples, Roger B., "Maritime Security Operations Mission Analysis Report Results and Actions," memorandum, Washington, D.C.: U.S. Coast Guard, December 21, 2014.

Ratti, S. H., "FY14 Operational Planning Direction (OPD)," memorandum to Coast Guard sectors and air stations, Portsmouth, Va.: U.S. Coast Guard, February 21, 2014, Not available to the general public.

Ryan, J. E., *Living Marine Resource Enforcement Performance Plan, FY 2011–2016*, Washington, D.C.: U.S. Coast Guard, July 1, 2010, Not available to the general public.

Servidio, Joseph, *U.S. Coast Guard Marine Transportation Systems Management Program Performance Plan, Fiscal Years 2014–2019*, Washington, D.C.: U.S. Coast Guard, August 2013a, Not available to the general public.

———, *U.S. Coast Guard Maritime Prevention Program Performance Plan, Fiscal Years 2014–2019*, Washington, D.C.: U.S. Coast Guard, August 2013b, Not available to the general public.

Shaw, D. P., "United States Coast Guard Readiness Reporting Resource and Capabilities Evaluation System (CG-RACE System)," briefing slides, Norfolk, Va.: USCG Force Readiness Command (FORCECOM), December 2014.

Sweigart, D. Blair, "Sectors 101," briefing slides, Portsmouth, Va.: U.S. Coast Guard, March 26, 2014.

Tambe, Milind, Will Haskell, and Matthew Brown, "USC and USCG: Lancha Interdiction," briefing slides, Los Angeles: National Center for Risk and Economic Analysis of Terrorism Events, 2014.

U.S. Air Force, "United States Air Fiorce Fiscal Year 2015 Budget Overview," Washington, D.C.: Air Force Financial Management and Comptroller, Deputy Assistant Secretary for Budget (SAF/FMB), March 2014. As of August 26, 2015:
http://www.saffm.hq.af.mil/shared/media/document/AFD-140304-043.pdf

USCG—*See* U.S. Coast Guard.

U.S. Coast Guard, "Atlantic Area Minimum Mission Standards," undated a.

———, "Fisheries Law Enforcement," in *U.S. Coast Guard Maritime Law Enforcement Manual (MLEM)*, Washington, D.C., undated b, pp. 9-1–9-18, Not available to the general public.

———, "Marine Protected Resources Law Enforcement," in *U.S. Coast Guard Maritime Law Enforcement Manual (MLEM)*, Washington, D.C., undated c, pp. 10-1–10-19, Not available to the general public.

———, *United States Coast Guard 2012 Posture Statement with 2013 Budget in Brief*, Washington, D.C., undated d.

———, *U.S. Coast Guard Strategic Blueprint*, undated e.

———, *Fisheries Enforcement Strategic Plan: Ocean Guardian*, Washington, D.C., September 2004.

———, "Seventeenth District Performance Plan," November 30, 2006.

———, *Ports, Waterways, and Coastal Security Performance Plan FY2011–2016*, 2010a, Not available to the general public.

———, *Drug Interdiction Mission Performance Plan FY2011–2016*, July 2010b, Not available to the general public.

———, *Marine Environmental Response Performance Plan FY2011–2016*, September 30, 2010c, Not available to the general public.

———, *Migrant Interdiction Mission Performance Plan FY2011–2016*, July 2010d, Not available to the general public.

———, "Coast Guard Snapshot 2012," flyer, Washington, D.C., 2012a. As of August 26, 2015: http://www.uscg.mil/top/about/doc/uscg_snapshot.pdf

———, *Command, Control, Communication, Computers, and Information Technology (C4&IT) Strategic Plan FY13–17*, November 2012b.

———, "Coast Guard Roles, Missions, Activities and Functions," in *America's 21st Century Coast Guard: Resourcing for Safety, Security and Stewardship*, Washington, D.C., 2013a, p. 4.

———, *Operational Sustainability Performance Plan*, Washington, D.C., February 2013b.

———, *Arctic Strategy*, Washington, D.C., May 2013c.

———, *Defense Operations Program Performance Plan FY2014–2019*, August 2013d, Not available to the general public.

———, *Maritime Law Enforcement Operations Program Performance Plan FY2014–2019*, August 2013e, Not available to the general public.

———, *Maritime Response Operations Program Performance Plan FY2014–2019*, August 2013f, Not available to the general public.

———, *Maritime Security Operations Program Performance Plan FY2014–2019*, August 2013g, Not available to the general public.

———, *Fisheries Enforcement Strategic Plan: Ocean Guardian*, Washington, D.C., 2014a.

———, *Always Ready: United States Coast Guard 2013 Performance Highlights [and] 2015 Budget in Brief*, Washington, D.C., March 2014b.

———, *Living Marine Resources: Ocean Steward*, Washington, D.C., September 2014c.

———, *Western Hemisphere Strategy*, Washington, D.C., September 2014d.

U.S. Coast Guard Atlantic Area, "LANTAREA FY14-15 Planning Assessment," May 2, 2013, Not available to the general public.

U.S. Code, Title 10, Armed Forces.

U.S. Code, Title 10, Armed Forces, Sec. 101, Definitions.

U.S. Code, Title 14, Coast Guard.

U.S. Code, Title 14, Coast Guard, Ch. 1, Establishment and Duties.

U.S. Code, Title 14, Coast Guard, Sec. 141, Cooperation with Other Agencies, States, Territories, and Political Subdivisions.

U.S. Department of Homeland Security, *U.S. Department of Homeland Security Annual Performance Report Fiscal Years 2012–2014*, Washington, D.C., undated.

———, *Small Vessel Security Strategy*, Washington, D.C., April 2008.

———, *Performance Measurement Verification and Validation Handbook: Determining the Completeness and Reliability of DHS Performance Measures*, Washington, D.C., September 2011.

U.S. Government Accountability Office, "GAO Executive Guide: Effectively Implementing Government Performance and Results Act," GAO/GGD-96-118, Washington, D.C., June 1996.

———, "Coast Guard: Strategy Needed for Setting and Monitoring Levels of Effort for All Missions," GAO-03-155, Washington, D.C., November 2002.

———, "Designing Evaluations: 2012 Revision," GAO-12-208G, Washington, D.C., January 2012.

U.S. Navy, "Status of the Navy," web page, August 26, 2015. As of August 26, 2015:
http://www.navy.mil/navydata/nav_legacy.asp?id=146

Watanabe, Todd, Unified Performance Logic Model (UPLM), briefing, Washington, D.C.: U.S. Coast Guard, undated.

Young, Stephanie, Henry H. Willis, Melinda Moore, and Jeffrey Engstrom, *Measuring Cooperative Biological Engagement Program (CBEP) Performance: Capacities, Capabilities, and Sustainability Enablers for Biorisk Management and Biosurveillance*, Santa Monica, Calif.: RAND Corporation, RR-660-OSD, 2014. As of August 5, 2015:
http://www.rand.org/pubs/research_reports/RR660.html